Nuclear Transmutation of Stable
and
Radioactive Isotopes in Biological Systems

Nuclear Transmutation of Stable and Radioactive Isotopes in Biological Systems

Vladimir I. Vysotskii
Professor, Head of Dept. of Theoretical Radiophysics
Radiophysical Faculty, Kiev National Shevchenko University
Kiev, Ukraine

Alla A. Kornilova
Ph.D.
Director of Innovation Centre, Physical Faculty
Senior Researcher at Solid State Physics Dept
Moscow State University, Moscow, Russia

Foreword by

B.M. Hegde

PENTAGON PRESS

Nuclear Transmutation of Stable and Radioactive Isotopes in Biological Systems /
Includes Foreword, Bibliography Vladimir I. Vysotskii, Alla A. Kornil

ISBN 978-81-8274-430-1

First Indian Edition 2010

Published by
PENTAGON PRESS
206, Peacock Lane, Shahpur Jat,
New Delhi-110049
Phones: 011-64706243, 26491568
Telefax: 011-26490600
email: rajan@pentagon-press.com
website: www.pentagon-press.com

Printed at
Ritomate International
Noida Special Economic Zone, India

Foreword

This book has come as a great relief for my long standing suffering. At the time when I was a graduate student of medicine in the 1950s at the Stanley Medical College, Madras, we had a textbook of clinical biochemistry by King. While we were lost in the concept of Claude Bernard's *milieu interior*, the constant internal environment, the book wrote that to replace potassium deficiency, the calculated amount should not be fully replaced but only half that amount should be replaced. That was very confusing. What happens to the other half to get back to the constant internal milieu? My professor rebuked me for asking that query. My concern continued, though. Later, after years of treating sick individuals, I realised that all is not well with our thinking in this area. Another area of great concern was fluid and electrolyte replacement. While the "wet theorists" advocated drop per drop replacement of fluids and blood, the "dry theorists" cautioned prudence as the human body has the capacity to replace quite a bit by *auto-transfusion* and *auto-infusion*, the two mechanisms that defied our reductionist logic in medical science.

Audits of per capita deaths in grievously injured soldiers in the Vietnam War and those of the Falklands War gave me a rude shock again to know that, despite the best emergency treatment facilities in Saigon minutes away from the war theatre, the British soldiers left in the snow covered ground for well over 24 hours and then transported to London were marginally lower. I was wondering if the body's internal corrective systems could manage most of the aberrations in injury and disease states to keep mankind alive, to the extent possible, as the maker of this wonderful wonder, the "human body," did not anticipate hi-tech reductionist science of medicine to save mankind. Isn't it true that the recent audits in the USA had shown the medical interventions and mistakes to be the leading cause of death there? Vysotskii's book gives me an inkling as to how the King's bio-chemistry fiasco and the secret of reduced deaths in less interfered soldiers could occur.

Biological transmutations, the subject matter of this great book, is all about the golden secret of man's existence on this planet for 900,000 years in 50,000 generations despite the efforts of the medical establishment in the last one hundred odd years with all kinds of interventions (intervene = go in between with malice—Webster's definition, one of them) to disturb the working of this dynamic machine, the human body, which follows the non-linear laws of the new science of Chaos; not the deterministic predictability laws of Isaac Newton. All biological systems must be grateful to the concept of transmutations for keeping them going. I feel but for that capacity, we would have been extinct long before the dinosaurs perished. Science is simply trying to understand the beauty of the working of this nature. In that sense, science becomes a simple curiosity to get at the truth. Unfortunately, technology has been trying to use science to dictate nature how to run its agenda. The truth, sadly, becomes a casualty in this game at times.

The book was difficult to decipher in the first reading for a non-physicist like me, but second reading brought home most of the key points in the book. I must admit that the book is written basically for non-physicists like me, although hard core conventional physicists could understand biological transmutations if they had an open mind.

Science must learn the secrets of Nature instead of trying to teach Nature a lesson or two. All that has to be known in science is known to Nature as otherwise, this universe cannot exist at all. If we are humble enough to understand this fact, we will be ideal scientists. Understanding biological transmutations in Nature would solve many of the mysteries in modern science. The ancient Indian system of *Ayurveda* did have many treatment modalities like *Bhasmas* that were made using metals like mercury, silver etc. Interestingly, the final product, if manufactured strictly according to the original scriptures, will not contain any trace of that metal. Recent revelations have shown that many of those metals are in nano or pico form in the *Bhasma*. That brings us back to alchemy, the time honoured beginning of modern chemistry, which was later condemned by modern chemists. In reality, as shown in this book, alchemy might not be what it is made out to be in "modern science."

It was Ralph Waldo Emerson who wrote years ago that "in every work of genius we recognize our own rejected thoughts; they come back to us with a certain alienated majesty." This applies to this book—a work of genius-which brings back to us our rejected thoughts with certain majesty. A little neglect now might result in great loss. As Benjamin Franklin showed how "for want of a nail eventually the war was lost"! This germinating seed of the new science of transmutation needs patronage and people across the globe need to try and replicate the work exposed here to show the real truth for the good of mankind. Now everything that was haunting me in managing patients, the uncertainties, seem to make some sense in the background of the possibility of the human system capable of correcting most, if not all, deviations from the normal to a great extent. It is only in the unlikely event of the body failing to correct the deviations do symptoms develop. When symptoms develop it is right time for humane doctors to step in to "cure rarely, comfort mostly, but console always" taking the help of the body's inbuilt capacity to correct the defect. Our interventions should not damage the immune system's efforts to help in every conceivable calamity.

January 8, 2009

B.M. Hegde
World Academy of Authentic Healing Sciences,
"Ganesh," Lower Bendur,
U'sity Northern Mangalore-575002
Professor (Affiliate) of Human Health
Colorado,
Former Professor of Cardiology (Visiting)
London University,
Former Vice Chancellor, Manipal
University, Manipal, India. Former Director-Professor
and Dean, Kasturba Medical College, Mangalore, India.

Preface

Among the "eternal problems" contemplated by the human civilization throughout its entire history, the problem of unity and transformation of the material world has always occupied a special place. Philosophers and alchemists, poets and physicists studied this problem. The search for "philosopher's stone", capable of transforming elements to get gold in the end continued even during the darkest periods in the history of human civilization. Often, martyrs of science were sent to sacrificial fires of inquisition rather than renounce their progressive ideas. Among all the different races of the creative community, there were those, who sincerely believed in the power of human intellect, while—as always—there were also adventurists and charlatans, ready to promise everything at once, acting according to the principle "demand brings supply". Gold was in high demand in all times.

That was the "golden childhood" period in science, which was golden not in the last instance due to a naive and childish conviction that a simple variation of a few possibilities will definitely lead to success. It would be enough to mix something with something else and then add another ingredient and the sought objective would be achieved.

But progress was unstoppable in its onward movement. It seemed that the nineteenth and, especially, twentieth century have put everything in their places. The world has become ever more familiar. With the discovery of the structure of atoms and nuclei, the problem of transformation of the matter became perfectly real. Science became totally pragmatic. Meanwhile, the new caste of keepers of the "purity of science" has emerged.

The lines were clearly drawn. Everything that doesn't conform to the "generally accepted" canons was proclaimed heresy (not much different from returning to the times of the inquisition!). For example, when experts in "hot" thermonuclear synthesis claim that low-temperature synthesis is impossible, that is accepted as the truth. Only the most outstanding scientists (of the Nobel Prize caliber, such as A.D. Sakharov, U. Schwinger or B. Josephson) were allowed to express their opinion about theoretical possibility of such processes and not be afraid of being accused of supporting "false science".

But there was also one difference from the inquisition. The canons of science could be openly discussed. It appeared immediately, that part of them is definitely true (for example, the laws of preservation of energy), while another part is mythologized and corresponds to a particular level of knowledge.

Previously "proven" statements, that superconductivity at the temperature of over 30 degrees Kelvin is principally impossible (this canon has been successfully dismissed), or another similarly "proven" claim about the impossibility of existence of stable nuclei beyond the limits of the "stability island" with $Z \approx 118$ (this claim will soon be also defeated (Controlled Nucleosynthesis, 2007; Adamenko, Vysotskii, 2004, 2006), are among such "canons".

In fact, science was divided between professional clans, which longed for monopoly in specific knowledge. Any encroachments by "outsiders" holding radical views were treated harshly.

Meanwhile, life went on. The number of experiments that couldn't fit the Procrustean bed maintained by the hierarchy was growing. These experiments could be ignored, up to a certain point, or explained away by uncontrollable initial conditions or mechanical faults (the best option), if not labeled outright falsification. In all truth, there were some deliberate falsifications. Alas—a human being is weak and imperfect!

However, it didn't determine the facts, which became the "golden fund". And when the sheer volume of these facts has lapped over the critical mass, they had to be evaluated and properly explained.

Such facts are numerous. Sometimes, they lay on the surface, and sometimes they are hidden deep. There are especially many such facts in the fringe areas, where different sciences are closely and unusually intertwined (for example, biology, physics, chemistry). On the grand scale, life itself, realized as a form of existence of protein objects, is a big mystery. It is an anti anthropic process, maintaining order in the world of chaos. Its explanation from the point of view of a big fluctuation cannot, essentially, explain anything.

Other phenomena occurring in living objects are not less mysterious.

Stability of DNA, giving the probability of mutation of no more than 10-7 regardless of local conditions and the make up of nutrient media, is one such mystery. This fantastic procreative consistency throughout the whole specter of external conditions cannot be explained by the effect of a DNA polymerase controlling that process. In a nutrient environment, some necessary chemical elements may be missing, external conditions may change, and yet the process of undistorted and non-mutating procreation prevails with constantly high efficiency! Even super powerful radiation fields cannot disrupt it. The bacterial culture *Deinococcus Radiodurans* comfortably lives, multiplies and does not mutate in radiation fields with the dose exceeding 10 Mrad! Many structural materials cannot sustain such dose of radiation (not to mention such materials as crystallized silicon, which is the basic material for all microelectronic devices, and which completely degrades in such conditions), but a biological object, maintained by the rules of genetic information can successfully live and develop! It appears that biological objects have an internal source of high stability and reliable self-regulation, which inanimate objects do not have. How does it happen? Where is that internal source of stability? What are the limits of self-regulation? Can these processes be controlled?

Part of these questions is discussed in this monograph, which also attempts to answer some of them.

This work is mainly dedicated to the analysis of mechanisms, capable of sustaining processes of nuclear transmutation of isotopes in growing biological systems. It is, in fact, an attempt to present a live object as a natural nuclear micro reactor, which can, with high efficiency, provide the process of transformation of nuclei in clearly "non-reactor" conditions (room temperature, normal atmospheric pressure, absence of any high energy particles).

Why can a living system, seemingly fragile and sensitive to external conditions, perform the process, which is practically impossible to achieve (at least with a comparable degree of efficiency) in inanimate objects? How does it do it? Why does it do it?

When answering these questions, we proceeded strictly from solid facts and methods of modern science. We didn't invent new myths, but based our deliberations on the principle of "not to create any substance in excess of the necessary" and tried to explain all facts using the tested methods, without introducing (as it is often done) a new radical theory to explain each fact. In this respect we are definite conservatives. For this reason, we don't think, that the totality of the effects being analyzed can be called "biological transmutation". That name would give it a semi-mystical flavor, which, by no means, reflects the essence. In reality, it is transmutation of nuclei (isotopes) in biological systems occurring according to strict laws of physics, but induced by certain features of biological objects' structure, functioning as stimulators and intrinsic in dynamic living systems.

In this work, the processes of transmutation of isotopes in growing biological systems are examined from three different angles—as totality of experimental facts of low temperature transmutation of isotopes, as a process based and analyzed from the perspective of nuclear science, and as a process studied from the point of view of biochemistry of live systems. To the extent of our understanding, we were trying to show that nature is one, and, therefore, it is pointless to separate and then compare processes in living dynamic matter with effects occurring in their static surroundings. On the other hand (and that was our main point), we wanted to demonstrate that only such dynamic and self-regulating living space could provide answers to many questions regarding evolution of the matter.

The present publication is a natural logic sequel of our preceding book (*Vysotskii, Kornilova, 2003*), where we considered the problem, some theoretical aspects and experements on nuclear fusion and transmutation of stable isotopes in biological cultures.

In the present book the problems of nuclear fusion and transmutation of both stable and radioactive isotopes in biological systems with presence of regular (normal), activated (with changed structure) and radioactive water are considered.

In conclusion, we would like to express our gratitude to our friends and colleagues from Kiev National Shevchenko University (Ukraine), Lomonosov Moscow State University (Russia), Gamaleya Institute of Epidemiology and Microbiology (Russia), Zabolotny Institute of Microbiology and Virology of the National Academy of Sciences of Ukraine (NASU), Kavetsky Institute of Experimental Pathology, Oncology, and Radiology of NASU (Ukraine), Kiev Institute for Nuclear Research of NASU (Ukraine), who participated in our research.

An active cooperation of Dr. I.I. Samoylenko, Dr. A.B. Tashyrev,

Dr. S.I. Reyman, Dr. G.A. Zykov, Dr. E.S. Vasilenko, Dr. A.V. Syroezhkin, Dr. A.M. Pokidyshev, Prof. V.N. Pavlovich, Dr. A. Odintsov, Dr. Yu.V. Yanish, Dr. S. Olishevsky, Prof. L.S.Kholodna, Prof. V.D.Rusov has made this book possible.

We would also like to thank the participants of the International annual ICCF conferences on “low energy nuclear reactions” (former “cold nuclear fusion” conferences) and Russian monthly seminar “Problems of the low-temperature nuclear synthesis” (and its organizers—Dr. N.V. Samsonenko and Dr. Y.N. Bazhutov) for their inspired discussions of the issue of isotope transmutation over the years.

Contents

Foreword v

Preface ix

Chapter 1

The Issue and the History of the Effect of Chemical Elements Transmutation in Biological Systems **1**

Chapter 2

The Isotopic Content and the Impact of Isotopes from Biosystems **10**

Chapter 3

Experimental Study of Transmutation of Stable and Radioactive Isotopes on Microbiological Cultures and Microbe Syntrophin Associations **16**

3.1. Choosing experimental methodology and biological objects used 16

3.2. Comprehensive study of controlled transmutation of isotopes in fast growing biological cultures using the Messbauer and mass-spectrometry methods 18

3.2.1. Study of isotopic transmutation in microbiological cultures using the Mossbauer spectrometry method 19

3.2.2. Study of controlled transmutation of light and intermediate isotopes in a growing microbiological culture using the laser time-of-flight mass-spectrometer 29

3.3. Monitoring and mass-spectrometry analysis of controlled transmutation of isotopes with middlerange atomic numbers in a growing culture 33

3.3.1. Experimental methodology and procedures 33

3.3.2. Results and discussion 36

3.4. Observing transmutation of heavy isotopes and studying methods of controlling transmutation processes in growing microbiological cultures 40

3.4.1. Catalytic effect of cesium on effectiveness of nuclear transformation of medium and heavy isotopes in a growing biological culture 40

3.4.2. Detection of transmutation of heavy isotopes in growing biological cultures 45
3.4.3. Study of the combined effect of Sr, Cl and S on efficiency of the process of transmutation in biological cultures 47
3.5. Effective nuclear fusion and transmutation of stable isotopes in microbe syntrophin associations 52
3.5.1. The general features of isotopes transmutation in "one-line" cultures and in microbiological associations 52
3.5.2. Experiments on transmutation of isotopes in growing microbiological associations 55
3.6. Experiments on controlled nuclear transmutation and deactivation of long-lived radioactive isotopes in biological cells 59
3.6.1. Accelerated deactivation of reactor water in biological cells 59
3.6.2. Accelerated deactivation of long-lived active Cs^{137} isotope in microbe syntrophin associations 63
3.7. Investigation of long-term memory phenomena in water and study of isotopes transmutation in growing biological systems containing activated water 66
3.7.1. The problem of long-term memory of water 66
3.7.2. Investigation of physical-molecular properties of activated water 70
3.7.2.1. Anomalous electrodynamic characteristics of activated water 71
3.7.2.2. Anomalous viscosity characteristics of activated water 74
3.7.2.3. Influence of the activation of water on pH 76
3.7.3. Investigation of inhibition biological activity of activated water 80
3.7.4. Initial experiments on transmutation of intermediate mass isotopes in growing biological cultures in activated water 85
3.7.5. Conclusions 85

Chapter 4

Physical Aspects of Low Temperature Transmutation of Isotopes in Biological Systems **88**

4.1. Low energy non-threshold nuclear reactions of synthesis 88
4.2. The phenomenon of "cold nuclear fusion" 92
4.2.1. Accelerating models 96
4.2.2. Reducing the Coulomb barrier through reflection of electrons 96
4.2.3. Coherent and interference effects involving lattice 96
4.2.4. Fermi and Bose nuclei condensation 96
4.2.5. Reactions stimulated by lattice phonons 97
4.2.6. Exotic chemistry 97
4.2.7. Catalysis by heavy (transmuonic) particles 97

4.2.8. Reactions under the influence of neutrons 97
4.2.9. Peculiar nuclear effects 97
4.2.10. Reactions under the influence of hypothetical magnetic monopoles 97
4.3. Distinctive features of producing the "cold nuclear synthesis" phenomenon in biological systems 98
4.3.1. Is there a distinctive mechanism of producing low temperature nuclear synthesis in biological systems? 98
4.3.2. Potential mechanism of non-barrier nuclear synthesis in optimal potential quantum wells in the non-stationary mode 101
4.3.3. Possible scenario of running barrier-less nuclear synthesis in growing biological systems 113

Chapter 5

Essential Biochemical Factors of the Process of Transmutation of Isotopes in Biological Systems 116

5.1. Elementary constitution of biological objects and the problem of controlled synthesis in a growing culture 116
5.2. Specific features of microorganisms' interaction with metals 121
5.2.1. Metals mobilization 122
5.2.2. Immobilizing of metals 123
5.2.3. Accumulation of metals 123
5.3. Biological significance of microorganisms' interaction with metals 126
5.4. Biochemical analogues of vital microelements, metabolism of microorganisms and the question of controlled transmutation 128
5.5. DNA restoration and the problem of transmutation of isotopes 139
5.6. Adaptation of biological objects, analogy of microelements and synthesis of isotopes –the rule of biological evolution 142

Chapter 6

Prospects and Possible Methods of Using the Phenomenon of Transmutation of Isotopes in Biological Systems 144

6.1. Using the phenomenon of transmutation of isotopes for prevention and treatment of illnesses, etiologically caused by deficit of microelements 144
6.2. Obtaining rare light and heavy stable isotopes by biotechnological methods, based on the phenomenon of transmutation of isotopes 148
6.3. On the possibility of radioactive waste utilization using the effect of transmutation of isotopes in biological systems 152
6.3.1. Reaction of utilization of long-living isotope Sr^{90} 161
6.3.2. Reaction of utilization of radioisotope Zr^{95} 162
6.3.3. Reaction of utilization of Nb^{95} 162

6.3.4. Reactions of utilization of highly active isotopes Ru^{103} and Ru^{106} 162
6.3.5. Reactions of utilization of highly active nuclei of Sb^{125} 163
6.3.6. Reactions of utilization of highly active isotope J^{131} 163
6.3.7. Reactions of utilization of highly active isotope Co^{60}, used in metrology and flaw detection 164
6.3.8. Utilization of highly active isotope Kr^{85} 165
6.3.9. Utilization of highly active isotope Sc^{46} 165
6.3.10. Utilization of highly active isotope Mn^{54} 166
6.3.11. Utilization of highly active isotope Ar^{39} 166
6.3.12. Utilization of highly active isotope S^{35} 166
6.4. Conclusion 167

Bibliography 169

Index 188

CHAPTER 1

The Issue and the Brief History of the Effect of Chemical Elements Transmutation in Biological Systems

The issue of transmutation and synthesis of chemical elements during the "pre-nuclear period" has its own history and mythology, its own proponents and critics. To a significant degree, this problem was based on the critical analysis of alchemy, which was a bizarre mix of the elements of rational empirical knowledge about chemical processes and traditional medieval mysticism. In the end, these elements have comprised a major component of modern chemistry. The fundamental laws of chemistry were formulated, which made the process of transmutation impossible. These laws, to a large extent, are related to preservation of certain invariables (among which energy is the most important one). Such condition has significantly narrowed the specter of display for the process of transmutation but has not disallowed it altogether.

It is worth noting, that the issue of transmutation of chemical elements in biological objects resided, mainly, outside the boundaries of interest for both the medieval and the "pre-nuclear" science. To a significant degree, it was due to theological mores of that time, juxtaposing animate objects to their inanimate environment.

That situation prevailed, practically unchanged, until the first third of the 20th century. With the discovery of the atomic structure and the nature of sub-atomic interaction, it became more obvious that, although all chemical features of the elements are caused almost exclusively by the behavior of valence electrons, still, the very nature of these elements (and, ultimately, the characteristics of valence and other electrons of the outer shell) are tied to the nuclear characteristics of the isotopes.

The series of works by Louis C. Kervran (Kervran, 1963, 1966, 1968, 1998) holds a special place in the chronology of transmutation of chemical elements and isotopes in biological objects. Effectively, he was the first scientist of the post-nuclear era, who conducted systematized research of possible transmutational processes of chemical elements in biological objects.

In his works Kervran gave numerous examples of unusual changes in the chemical composition of various biological objects, which occurs during their growth. He explained these changes on the basis of existing concepts of transmutation of chemical elements.

For example, among the most important, he investigated the reaction

$$K^{39} + p^{1} = Ca^{40}$$

of potassium transmutation into calcium in the biological system containing hydrogen (with a participating proton). His works contain reasonably convincing experimental results, illustrating this process (see Fig. 1.1). This data corresponds to changes in potassium and calcium content in the process of growing seeds and were obtained from the analysis of 840 seeds and 403 sprouts.

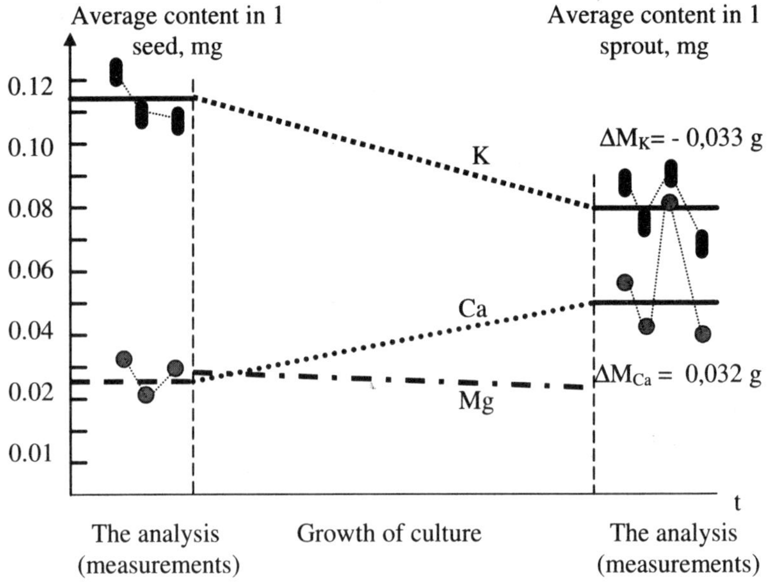

Fig. 1.1. Changes in K and Ca content in the seeds and sprouts. The left and the right parts of the figure show the measurement results by three series and average results

It can be seen that the decrease in the amount of potassium $(\Delta M_K = -0.033\, g)$ and the increase in the amount of calcium $(\Delta M_{Ca} = 0.032\, g)$ are approximately equal.

Meanwhile, the amount of magnesium stays practically unchanged.

Kervran also investigated many other reactions of transmutation of isotopes, among which several should be specifically noted for their vital activity in producing essential elements Ca, K, Mg, P

$$Mg^{24} + O^{16} = Ca^{40},$$
$$Si^{28} + C^{12} = Ca^{40},$$
$$Na^{23} + O^{16} = K^{39},$$
$$Na^{23} + p^{1} = Mg^{24},$$
$$Mg^{25} + Li^{6} = P^{31},$$

as well as the reaction of transmutation within the molecule $N_2 \rightarrow CO$, corresponding to the two-stage transformation of isotopes

$$N^{14} + N^{14} \rightarrow C^{12} + n^{1} + p^{1} + N^{14} \rightarrow C^{12} + O^{16}$$

He considered the latter reaction as the process of proton and neutron transformation in the N_2 molecule from one nucleus of nitrogen to another (with transformation of one nucleus of nitrogen into carbon and another—into oxygen) and suggested that this process will take place in a biological system in conditions of carbon deficit.

As a proof of running the reaction of synthesis $Na^{23} + O^{16} = K^{39}$, Kervran provided the experimental data (Jullien, 1959). According to this data, placing a tench fish into water, containing 1.4% of sodium chloride NaCl for 4 hours, a 36% increase in the general KCl concentration, from the initial amount of 3.95 g/L to 5.40 g/L, was observed. Changes in content of NaCl and KCl in the blood of a tench fish for the same period of time are presented on Fig. 1.2.

Such examples, including those from the area of botanic, applied agriculture, bird farming, are abound in his works.

Special attention is given, in his works, to the possibility of using the phenomenon of transmutation in biological systems for radioactive waste disposal involving transmutation of unstable elements into stable ones.

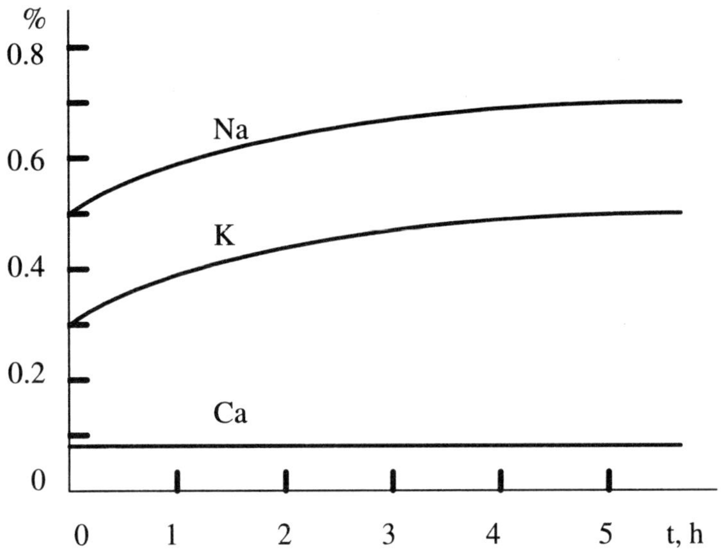

Fig. 1.2. Changes in content of Na, K and Ca in the blood of a tench fish, immersed in water containing 1.4% of NaCl.

As a possible example of such process, Kervran offers the process of the microbiological culture *Klebsiella aerogenes* development in the media, containing a radioactive isotope Hg^{203} with a half life period of 46.6 days in the form of chemical compound $HgCl_2$. The results of these experiments (Magos, 1964) have shown that the rate of radioactivity reduction in the media, containing active cultures, during 16 hours, was higher than under normal conditions of decay. A standard Geiger counter measures radioactivity. Although the authors related the fact of divergence in radioactivity reduction rate from the normal rate of decay to evaporation of radioactive mercury, Kervran suggested, that in this case there is an effect of transmutation of the radioactive isotope Hg^{203} into non-radioactive nuclei.

In summary, it can be noted, that Kervran analyzed possible ways of transforming elements and isotopes using common sense algebra rules for nucleons—the unchanged number of protons and neutrons in these transformations.

This approach, notwithstanding possible accompanying mutual transformation of the nucleons considering weak interaction, does not raise major objectives. Moreover, Kervran's perception of the nuclei structure and possible ways of their transformation were very different from customary and contemporary views in nuclear physics. More specifically, considering the reaction analyzed above,

$$N^{14} + N^{14} \rightarrow C^{12} + n^{1} + p^{1} + N^{14} \rightarrow C^{12} + O^{16},$$

he considered the process of the proton's transformation into a neutron in the N_2 molecule from one nucleus of nitrogen to another, as taking place on the cell's surface or in the volume of the membrane given the catalyzing power effect of an unknown enzyme.

It is a standard practice in biochemistry—to attribute any new and previously unknown phenomenon to a new, previously unknown enzyme. This approach—postulating a new enzyme—in fact, solves nothing and neither gives an answer, but creates an illusion of a scientific approach. However, in this case, such method is clearly unacceptable because an effect of any enzyme is manifested through its influence on outer electrons and matches the scale of the power interaction on the level, equal to 0.1–0.5 eV. Nuclear transformations require power interactions to be over 100 KeV!

It is also worth noting, that, although Kervran understood the significance and methodology of investigating transmutation of isotopes specifically, as opposed to chemical elements in general, his works demonstrated that his strict adherence to methodology was often too formal—a tribute to the requirements of proper recording rather than a reflection of the essence of the process of nuclear transformations.

Formalized attitude towards the methodology of the isotopic content becomes evident in recalling his often used concept of the reversibility of the transmutation process, which allowed him—without any credible justifications to postulate the unthinkable for nuclear physics process of reversible threshold nuclear reactions, which violates the law of preservation of energy.

For example, he postulated the possibility of reversing the reaction of potassium transmutation into calcium

$$K^{39} + p^{1} \leftrightarrow Ca^{40}$$

in changing external conditions (i.e. inversion of potassium and calcium concentration in the environment).

However, the reversibility of these processes is incompatible with the law of preservation of energy. If, in the case of the direct reaction of Ca^{40} synthesis, there is emission of energy. ΔE = 8.326 MeV per each transmutation (it is non-threshold and energetically productive) and it does not contradict the law of preservation of energy, the reverse reaction must have thresholds and can run only if the initial nucleus Ca^{40} will receive energy, not less than the same 8.326 MeV. The probability of a live culture, existing and developing in the temperatures $T \approx 300 - 340\,K$, receiving, by the way of fluctuations, such additional energy, is equal to a very small number

$$W \approx e^{-\Delta E/kT} \approx e^{-100000000}.$$

With this probability, there would be no single incident in the history of the Universe! Obviously, such reversible reaction is practically impossible, particularly if such a reversed reaction will be taking place simultaneously in many nuclei.

Such examples of careless assumptions are numerous in Kervran's works.

For instance, reactions of direct fission of isotopes, analyzed by him,

$$Cl - O \rightarrow F,$$
$$P - Li \rightarrow Mg,$$
$$Ca - O \rightarrow Mg,$$
$$Fe - p \rightarrow Mn,$$

which, according to his opinion, can be sustained in living systems, are exothermal and with thresholds and can be run in a controlled direction with decreasing atomic number only with a huge amount of additional energy, equal to 5–20 MeV for a single reaction. It is impossible to attribute such process to the effect of an unknown enzyme. No such source of free energy exists in living systems!

Similar arguments can be presented relating to the Kervran's interpretation of the process of radioactivity reduction in the environment, containing radioactive isotope Hg^{203}. This interpretation was based on the registered deviation from the law of natural decay observed for 16 hours in an experiment (Magos, 1994). The specter of natural decay of such a nucleus corresponded to the emission of fast neutrons with energy, equal to 214 KeV and gamma-quants with energy, equal to 279 KeV. If the Kervran's interpretation of quick transmutation of Hg^{203} into a non-radioactive isotope is accepted, then the most logical suggestion would be that isotopes with the highest concentration in the environment should participate in the process of transmutation, that is, the hydrogen nucleus p^1 and oxygen O^{16}. The result of these reactions is the creation of an unstable isotope Tl^{204}

$$Hg^{203} + p^1 = Tl^{204},$$

with a longer life-time (τ= 3.78 years) and initiating a chain of nuclear transformations, accompanied by emission of alpha-particles

$$Hg^{203} + O^{16} \rightarrow Ra^{219}(\tau = 0.01\sec,\ He^{4}) \rightarrow \mathrm{Re}^{215}(\tau = 0.002\sec,\ He^{4}) \rightarrow$$
$$Po^{211}(\tau = 0.5\sec,\ He^{4}) \rightarrow Pb^{207}$$

The final result of these reactions is formation of stable isotope of lead Pb^{207}.

These reactions could, indeed, lead to a reduction of radioactivity, but their execution in the conditions of this experiment is impossible. All these reactions have thresholds and require substantial additional energy (8.2 MeV in the case of the reaction with protons and 39.4 MeV for the reaction with oxygen) for their execution. The system does not have such additional energy and, consequently, such reactions are principally not feasible.

Apparently, another assumption seems more plausible—that the reduction of the counting rate at some moments of the cultures' growth is associated with a possible redistribution of the growing culture in space during its growth, the non-stationary non-homogeneity given the culture's growth, and other space-time factors. For example, increasing the weight of the growing culture in the near surface layer of the dish leads to a natural reduction of $HgCl_2$ concentration in the same volume (this salt will be pushed out to other parts of the volume) and increases the distance, that the fast electrons from products of Hg^{203} decay must cross in order to reach the surface, beyond which there is the Geiger counter. These processes naturally cause the slowing of fast neutrons emitted by the isotope of Hg^{203} into the volume of the media and, eventually, affect the stream of these electrons beyond the limits of the media. A significant slowing down of the electrons in the water results in a situation, when, in fact, only the electrons emitted by sources close to the near-surface layer, are registered. Provided that using the Geiger counter does not allow distinguishing between detection of quanta and fast neutrons, it would be regarded as a reduction of activity of the media. After the culture's growth is finished (due to poisoning of the nutrient medium by metabolism products, for example), there can be natural sagging or homogenizing of the grown culture throughout the entire volume of the media, which should immediately lead to a redistribution of $HgCl_2$ throughout the volume and a natural increase of its concentration in the sub-surface layer. This will result in a relative increase of the rate of detection of fast electrons and restoration of the law of normal decay. Naturally, this is only one of the possible explanations for the observed effect of non-exponential normal decay, yet it shows that without conducting controlling experiments and considering alternative mechanisms, Kervran did not have solid foundations for interpreting the results of these measurements as corroborating the existence of the effect of utilization of radioactive isotopes with their transformation into non-radioactive ones.

The effect could be described more definitely after carrying out an analysis of the isotopic content of the culture or the nutrient medium after stopping its growth, although all, without an exception, experimental data found in the works of Kervran, was obtained by chemical (elementary) analysis. In many experiments on transmutation of stable or radioactive isotopes conducted and analyzed by Kervran, there hasn't been a single case of performing the isotopic analysis of the products of transmutation, although there can be no reactions of transmutation of the nuclei apart from those, which involve transformation of isotopes! In fact, the very concept of a chemical element is just a convenient logical casing, holding an array of different isotopes. While, in some cases, such assumption can be justified, in other situation it is absolutely inadmissible!

It is particularly relevant for nuclear-physical processes, when features of different isotopes of the same chemical element are as similar as isotope features belonging to different chemical elements! However, none of the experiments and cases, analyzed by Kervran, have the analysis of the initial (before the start of the reaction) and the final (after its completion) isotopic content, although without such an analysis it is useless to speak about any reaction of transmutation. All chemical analytical methods provide information only about the elementary composition, while it is always the isotopic content that is always altered in a reaction of transmutation!

The direct continuation of Kervran's experiments were the works of his disciple Komaki (Komaki, 1965, 1967, 1993), dedicated to the study of processes of transmutation of chemical elements in growing microbiological cultures. The same comments that were made about the works of Kervran are fully applicable to these works as well. Essentially, these works were based on the transmutation of isotopes, but in none of these experiments there has been an analysis of the initial (before the start of the reaction) and the final (after the end of the reaction) isotopic content, and studies were limited to measurements of changes in chemical elementary composition.

One more comment is still in order associated with the very name for the phenomena of transmutation, which Kervran uses. In all of his works, he calls the process of transformation of elements *biological transmutation*. It was the direct result of his understanding of the essence of transmutation processes. He claimed, in particular, that specific biological rules may not be directly converted into the rules of physics, because all physical laws were obtained based on experiments with inanimate objects.

In our opinion, there are no reasons to consider the process of transformation of isotopes and elements biological transmutation and separate it from the general physical concept of transmutation as a process of transformation of isotopes, governed by the laws of physics. We believe that all the observed isotopic effects (in case they are real and supported by adequate and reliable measurements) can be characterized as the "regular" process of transmutation of isotopes and elements, which occurs in biological systems, and the efficacy of which is determined precisely by the specifics of such systems. The question of whether there is a certain quality intrinsic to biological objects, which stimulates the flow of nuclear processes, will be considered more rigorously later in the chapter discussing the theoretical aspects of the process of transmutation.

That chapter will also discuss the question of which mechanism is capable of providing conditions for the flow of transmutation reactions of synthesis in growing microbiological cultures.

CHAPTER 2

The Isotopic Content and the Impact of Isotopes from Biosystems

It follows, from the previous discussion, that the process of transmutation of chemical elements in biological systems is, primarily, the process of nuclear transformation of isotopes. Several questions arise with regard to it:

Are isotopes of any specific importance in the life cycle of biosystems?

Can the isotopic content in a biosystem be different from its natural level?

These questions should be discussed not only from the point of view of the general questions of metabolism, but also because of more pragmatic reasons, linked to the analysis of the processes of transmutation of isotopes based on detailed investigation of changes in the isotopic content. Obviously, reliable registration of transmutation processes is possible only when these changes sufficiently exceed natural variations of the isotopic content, caused by their differing physical and chemical characteristics. It appears, at the first glance, that since vital activity is mostly an aggregation of biochemical processes involving valence electrons, which are not significantly affected by the isotopic content of the nucleus, the isotopic content of a specific chemical element should not have a significant impact on the vital activity of organisms and all isotopes belonging to the same chemical element should behave in a similar manner. This fact—seemingly self-evident—is the foundation of the method of isotopic indicators, among which the method of using radioactive isotopes—chemical analogues of the stable isotopes of the same chemical element—as "labeled atoms", is the most prominent one. Based on these theses, it could be expected, that the ratio of isotopes in animate objects should correspond to the similar ratio in chemical elements of non-bionic origin. However, direct experiments show significant deviation of the isotopic ratio in live organisms and a strong influence of the isotopic content on a range of vital processes. A range of main mechanisms, potentially causing isotopic anomalies, can be identified. Some of these mechanisms were studied in detail in the book (Galimov, 1981, in Russian).

1. The kinetic isotopic effect, explained by different speeds of atoms (molecules) of different isotopes with the same thermal energy. Other effects, at- tributed to the fact that the chemical links formed with the participation of a heavier isotope are stronger than the links formed with the participation of a lighter isotope, are also related to this mechanism. The latter effect has simple quant/mechanical substantiation and explained by the fact that the energy levels of any heavier particle in a potential pit (characterizing that particle's interaction with its neighbors) are lower than the levels, defining the condition of a lighter particle. Factually, this is due to the stability of the minimal portion (quant) of the phase volume

$$\Delta\Gamma = \int d^3x \int d^3p = (2\pi\hbar)^3,$$

where $p = MV$ (the particle's impulse).

Apparently, with an increase of the particle's mass (with other conditions held constant), the area of its localization predictably decreases and $\int d^3x \sim 1/M$, which allows the particle to be closer to the bottom of the potential pit. This circumstance leads to an increase of the activation energy of heavier particles E^* and reduces the speed of the reaction

$$v/v^* = \sqrt{\frac{M^*}{M}} \exp\left(-\frac{E-E^*}{kT}\right)$$

based on thermally induced processes. Due to such a relation, the products of the reaction are enriched by the light isotopes, while the residuum, which did not participate in the reaction is enriched by the heavy ones.

2. The thermodynamic isotopic effect. It is due to the fact that the presence of heavier isotopes leaves less free energy in the system than in the case of lighter isotopes. To a significant degree, this effect, as well as the previous one, is attributed to lower energy levels of heavier isotopes in the general power field of the system.

3. The nuclear/spin isotopic effect is caused by the peculiarities of spin-dependent effects, i.e. ultimately with the Pauli effect. The main feature of these effects is that, although the precise mechanism of atomic

and molecular interaction is linked to the external (and, usually, valence) electrons, the specifics of the quant statistics (including, for example, the nucleus and the electrons) suggests that the sum of the spins of all the electrons and nuclei is the determining factor. Therefore, changing the spin of the nucleus (which can occur with different isotopes) directly and significantly affects the statistics, distribution, and interaction of the atoms and molecules.

4. Space-adjusting effects, described by the fact, that minor changes in the spatial structure of atoms and molecules, associated with the isotopic changes of the nucleus mass, may lead to large aggregate effects, given periodically arranged systems of such atoms and molecules. Such effects take place, for example, in the case of periodically arranged large biomolecules' interaction with heavy or light water, which will be discussed later. In biology, plenty attention is given to variations of isotopic ratio for several most important elements—carbon, oxygen, hydrogen, nitrogen, sulfur (example: Galimov, 1981, in Russian). For instance, natural variations of the isotopic ratio C^{12}/C^{13} for carbon

$$\delta_c = \frac{(C^{12}/C^{13}) - (C^{12}/C^{13})_{\text{standard}}}{(C^{12}/C^{13})_{\text{standard}}},$$

can reach 12%, while such variations in live organisms lie in the range of 1.5– 3.5%. Fossilized remains of the Belemnitella americana, an arch structure dated to the Cretaceous period, has been chosen as a standard for the isotopic ratio (C^{12}/C^{13}) standard. It can be shown, for comparison, that the isotopic content of the atmospheric carbon dioxide is equal to $\delta_C = -0.7\%$, and residual carbonates are known to have natural variations in the range from $\delta_C = -0.5\%$ to $\delta_C = 0.5\%$. The range of ˜ for live organisms will vary in different biochemical components. For example, lipids are enriched with light isotopes (for them, ˜C lies in the range from $\delta_C = -0.8\%$ to $\delta_C = -3\%$) and for proteins and carbonates the value of δ_C on average equals the value of δ_C for the whole biomass.

The standard for the isotopic ratio of hydrogen and oxygen is the average isotopic content of the World Ocean (AICO), for nitrogen isotopes—atmospheric nitrogen, and for sulfur—meteorite troilite sulfur from the Devil's Canyon (MDC).

The isotopic content of the two main isotopes of oxygen in the atmospheric carbon dioxide

$$\delta_O = \frac{(O^{16}/O^{18}) - (O^{16}/O^{18})_{\text{standard}}}{(O^{16}/O^{18})_{\text{standard}}},$$

is equal $\delta_o = 4.1\%$. Atmospheric oxygen is enriched with a heavy isotope O^{18} and $\delta_o = 2.4\%$, which is above the ratio for sea water, which is, on average, $\delta_o = 0$. In water, oxygen in the carbonic acid is in the state of isotopic thermo dynamical balance with oxygen in the water. The isotopic content of sea plants is $\delta_o = 1.4–2.8\%$. The isotopic content of plants on the ground depends significantly on the temperature of the environment.

With high external temperature, live organisms are enriched with heavy isotopes, as a result of water evaporation. In general, ground plants have $\delta_o = 1.5–3.5\%$.

The isotopic effect is manifested specifically in hydrogen. It is known that the uniqueness of stable isotopes of hydrogen (H and D) is that their weights' ratio is two to one. For comparison, each of the rest of all significant biological elements has stable isotopes with differences in weights of only about 1–7%.

Due to such a big difference in the case of hydrogen, the degree of separation of isotopes and their natural fractioning can be very large. Moreover, these isotopes will behave differently in biochemical reactions. This difference becomes especially well realized in the analysis of the effect on biomolecules of light and heavy water. Currently, it is estimated, that water has the tri-dimensional tetrahedral arrangement of hydrogen links and identical to the structure of ice. The main difference of the liquid water structure from ice is that part of these links is ruptured under the impact of thermal fluctuations of the grid. While one mole of water molecules in ice corresponds to two moles of hydrogen links, the infrared spectral analysis data shows, that melting ice causes 46% of the links to break up (Buijs, 1963). Moreover, the number of molecules with two links is only 33% of all molecules, with one link—41%, and with no links at all—27% (in this case, the protons of one molecule are linked with surrounding molecules). Breaking up of hydrogen links leads to filling void spaces in the tetrahedral structure of ice with water molecules. The molecules are now "packed" more tightly, which leads to a higher coordination number.

The cooperative mechanism of the creation of hydrogen links between water molecules defines a very important quality of water, putting it apart from other liquids—the associative character of its structure. Thus, in room temperature, more than half of the links in water are ruptured. However, an even distribution of ruptured links among the molecules is not advantageous because of the cooperative nature of their formation. The process of links' completion corresponds to the minimum of energy, which causes formation—in micro areas—of shared associations with the maximum number of links per molecule and, therefore, having the structure similar to the structure of ice with the maximum number of links. Such associations are, in reality, unstable. Local energy fluctuations lead to their decomposition and appearance, by fluctuation, of new ordered associations—clusters. Clusters have the life span of about 10^{-10} sec, which corresponds to about 1000 molecular fluctuations.

The "flickering clusters" model, first investigated by Frank and Ben (Frank, 1957, 1958), was later developed by Nemethy and Scheraga (Nemethy, 1962), and currently is, perhaps, the most developed model of water.

Apparently, the parameters of the tri-dimensional tetrahedral arrangement of hydrogen links (as the parameters of the "flickering clusters") strongly depend on the mass and dimensions of water molecules, which differ in the case of "light" (H_2O) and "heavy" (D_2O) water. In a system, allowing for DNA interaction with such a quasi crystalline water, the regular (light) water structure is ideally compatible with the DNA structure. It was the result of the evolutionary process. Removing water causes the destruction of even such a rigid structure as the double spiral of DNA. It appears that this problem was first studied in detail in the work of Sutherland (Sutherland, 1957). The structure of a dehydrated molecule of DNA is very similar to the structure of a denatured one, but it is capable of full restoration at 75% humidity, when the sorbed water content is 50% of the dry weight, i.e. there are 10 water molecules per nucleotide.

Substituting regular (light) water for heavy water brings incompatibility between spatial periods of the quasi crystalline structure of water and the structure of the DNA surface, which immediately causes generation of large longitudinal (placed along the DNA axis) deforming tensions and alters the longitudinal DNA period.

Such deformations are especially important during DNA synthesis, because the DNA polymerase controlling the DNA synthesis process registers the appearing deformed complimentary nucleotide pairs as defective (mutant), which causes their elimination. These events would result in slowing of the replication process and creation of the mutant nucleotide pairs. Apparently, it is precisely this mechanism of spatial incompatibility of the periodical DNA structure and the periodical structure of the heavy water, that constitutes one of the main factors responsible for the isotopic effect, when hydrogen is replaced by deuterium in water. With a higher atomic number and charge of a chemical element, the relative difference in the influence of the isotopic content on biochemical processes decreases, which means that for medium and heavy elements the isotopic ratio in an inanimate environment and living objects becomes practically identical.

Therefore, it can be stated, that deviation of the isotopic ratio for light elements in the range of

$$\delta_X = \frac{(X^{A+\Delta A} / X^A) - (X^{A+\Delta A} / X^A)_{\text{standard}}}{(X^{A+\Delta A} / X^A)_{\text{standard}}} \approx \pm 3 - 5\%$$

from the "standard" value (for which $\delta_x = 0$) is quite plausible, which may be due to processes of natural enrichment or depletion of a certain chemical element of its own isotopes, and does not necessarily points at nuclear transmutations occurring in the system. Clearly, the question of pure nuclear nature of changes in the isotopic ratio (the question of the existence of nuclear transmutation) can become relevant only if deviations in the isotopic ratio in the system of a chemical element exceed 10–20%.

CHAPTER 3

Experimental Study of Transmutation of Isotopes and Chemical Elements in Microbiological Systems

3.1 Choosing experimental methodology and biological objects used

The hypothesis about the possibility of nuclear transmutation of chemical elements and their isotopes in physical, biological and geological systems with low energy of relative movement of interacting nuclei has been frequently discussed during the last decades. Interest toward this issue grew after systematic study of the phenomenon of cold nuclear fusion (CNF) based on dd-reactions in solid bodies has begun, although none of the experiments referred to earlier gave completely reliable results, which would prove the possibility of the CNF. In our minds, this is due to an incorrect choice of methodology and the subject of study. We believe that the right methodology allows to fulfill the maximum number of conditions for stimulation and detection of transmutation:

(1) the isotope obtained as a result of transmutation should not initially (before experiments start) be placed in an environment, where nuclear CNF-type transformations take place (or its initial content as an admixture is controlled with high precision in initial and final materials);
(2) the isotope being synthesized must accumulate in the environment (which requires that it doesn't correspond with the chemical element in the environment);
(3) the nuclei of the synthesized isotope should possess the characteristics (spin, intra-atomic resonance passages, magnetic moment), which makes them substantially different from neighboring (by their mass numbers) admixed isotopes of the same or surrounding nuclei;
(4) the method of detection should be characterized, on the one hand, by maximal sensitivity and, on the other hand, be capable

of providing reliable recognition of the isotopes generated in the course of the anticipated nuclear reaction against the background of chemically similar admixed elements and isotopes, i.e. to have distinguished resonance features.

Most of the experiments on detecting the dd-CNF based on detecting neutrons, tritium, or thermal effect in nuclear synthesis reactions with participation of the lightest isotopes conducted earlier did not conform to these requirements

$$d^2 + d^2 = He^3 + n^1,$$

$$d^2 + d^2 = T^3 + p^1$$

The intensity of the registered neutron radiation in these reactions did not exceed the background level caused by cosmic radiation. Accordingly, the expected accumulation of tritium did not exceed its initial content in heavy water D_2O, and regular electro-chemical processes mostly explained the observed thermal effects.

Experiments on transmutation of light chemical elements in biological cultures (where the initial and the final chemical and isotopic content and the admixtures concentration were not controlled) conducted by Komaki also do not match these requirements.

It is also obvious, that each separate method of detection cannot guarantee an absolute reliability of the registration process. For example, the isotopic mass-spectrometry method has high sensitivity, but can occasionally produce ambiguities in the interpretation of experimental rcsults—for instancc, the mass specter of an ion of a heavy isotope is little different from the mass spectrum of an ion of a multi-atomic molecule with the same mass number.

The alternative Mossbauer method of resonance gamma-spectrometry always identifies an isotope correctly, but it can only be used for identification of a limited number of isotopes.

Better reliability and accuracy of measurements can be achieved only by using maximally optimized spectrometry equipment (in particular, laser massspectrometers, which always get dissociation of molecules to the condition of separate atomic ions due to a higher plasma temperature) and by using several different types of spectro-meters.

Selecting the experimental body also entails dealing with optimiza-tion conditions for the process of low-temperature transmutation. Naturally, the biological experimental object should be characterized by a high rate of growth (gaining mass), allow for the possibility of running a large number of independent controlling experiments and be relatively well adapted to temperature and chemical makeup of the environment. Moreover, keeping in mind subsequent optimization, it is desired that it should not be sensitive to radiation, i.e. external and internal radiation by ionizing fields of various nature (gamma-radiation, fast electrons, neutrons and so forth).

Clearly, only the microbiological objects, which are absolute record holders in adaptability to external conditions, can meet all of these requirements.

For example, such a unique culture as *Deinococcus radiodurans* can sustain radiation of 1 million Rad, which is unsurpassed by any other live object and makes it a prime candidate for transmutation of radioactive isotopes.

Choosing the type of the expected reaction and the final isotope is also a matter of optimization. There are many attractive synthesis reactions, which do not have thresholds and can be run with low energy and be well reconciled with various limitations stemming from the specifics of a particular biological object. To a large degree, their selection is a matter of researcher's intuition and his understanding of the problem. One must also take into account limited time and resources. Based on these aspects, we have come to the conclusion that the optimal way (from the point of view of maximal accuracy and possibility of further optimization) is an overall study, on the first stage, of such a combination of reactions, which, by different ways, generate isotopes of the iron group. This task is achieved by this work due to the fact that the authors are well familiar with the Mossbauer effect and the technique of Mossbauer spectrometry, which is based (by 60–80%) on one of the isotopes of iron. Meanwhile, other reactions, which run simultaneously and independently of the reaction of iron synthesis in the same cultures, are also examined in this work.

3.2 Comprehensive study of controlled transmutation of isotopes in fast growing biological cultures using the messbauer and massspectrometry methods

The objective of conducted experiments was detection and study of the possibility of running a low-temperature transmutation of isotopes in growing microbiological cultures. Having acknowledged the need for simple registration, we made the decision to use the Messabauer effect on the basis of isotope Fe^{57} in our first experiments.

Iron is the integral part of the majority of live organisms. Importantly, for most biological objects, the "regular" isotope Fe^{56} and the rare Mossbauer isotope Fe^{57} are identical. From the point of view of nuclear physics, they are completely different nuclei.

The first series of experiments on biological transmutation of isotopes were carried out provided the possibility of investigating the results of the transmutation by several independent methods.

Those experiments were based on the expected synthesis reaction of the Mossbauer isotope Fe^{57}

$$Mn^{54} + d^{2} = Fe^{57}$$

in a microbiological culture, that grows in the iron-poor water-salt nutrient medium based on the heavy water D_2O containing manganese salt. Among the undisputable advantages of using manganese is its single stable isotope Mn^{55}.

This circumstance makes interpretation of experimental results unambiguous.

The result of the expected synthesis reaction is formation of a rare stable isotope Fe^{57}, concentration of which in natural iron is very small and equals 2.2%.

The apparent advantage of this reaction is because Fe^{57} is the most studied Mossbauer isotope, it can be easily identified using the Mossbauer effect. Accumulation of this isotope with increasing the reaction's efficiency and its duration makes possible (at least in perspective) to study temporary patterns of transmutation process. Above that, since natural concentration of Fe^{57} isotope is small, its synthesis can be easily registered by traditional mass-spectrometry based on the analysis of changes in relative distribution of isotopes of iron.

3.2.1 *Study of isotopic transmutation in microbiological cultures using the Mossbauer spectrometry method*

Experiments were conducted using several bacterial cultures (*Bacillus subtilis GSY 228, Escherichia coli K-1, Deinococcus radiodurans M-1*) as well as the yeast culture *Saccharomyces cerevisiae T-8*. Selection of these cultures was motivated either by their experimentally proven ability to grow in the heavy water based media or by the prospect of using the radiation-stable culture *Deinococcus radiodurans M-1* in transmutation processes given the presence of powerful radioactive fields, as was noted earlier.

A typical series of experiments included several consecutive steps.

The culture, preliminary treated in a centrifuge, thoroughly rinsed in distilled light water (H_2O), and being bred in a stimulating environment, was placed in a dish with controlled salt-sugar nutrient medium, containing basic (present both in trial experiments and experiments on transmutation) salts of Mg, Ca, K, ammonium tartrate, sucrose, and 10 ml of pure water (heavy water D_2O in experiments on transmutation, and light water H_2O in trial experiments). The list of ingredients of the nutrient medium is presented in Table 3.1. In order to eliminate (reduce) the influence of uncontrolled admixtures in the nutrient medium, reagents belonging to a chemically pure category, were used.

Table 3.1

Ingredients of the nutrient medium	*Concentration in the medium (%)*	*Fe admixtures (no more than) relative (%)*	*Fe admixtures (no more than) absolute (g)*
Sucrose	3	10^{-4}	3.10^{-7}
$(NH_4)_2$ tartrate	1	5.10^{-4}	5.10^{-7}
$MgSO_4 \cdot 7H_2O$	0,25	2.10^{-4}	5.10^{-8}
$CaHPO_4 \cdot 7H_2O$	0,008	$1,5.10^{-3}$	$1,2.10^{-8}$
K_3PO_4	0,5	5.10^{-4}	$2,5.10^{-7}$
$MnSO_4 \cdot 7H_2O$ (transmutation experiments)	0,01	5.10^{-4}	5.10^{-9}
Pure water (D_2O for transmutation experiments or H_2O for trial experiments))	100 (10 ml)	10^{-7}	10^{-8}

Possible Fe (as a chemical element) content for such a category in the volume of the nutrient medium did not exceed $\Delta m_{Fe} \approx 1.1\times10^{-6}$ g in a single experiment. According to the natural concentration of the isotope Fe^{57} (2.2% of the total amount of admixed iron), its possible amount in the form of admixture did not exceed $\delta m_{Fe\text{-}57} \approx 2.5\times10^{-8}$ $\tilde{g}$ (Table 3.1).

In transmutation experiments in the nutrient medium based on heavy water, salt $MnSO_4$ containing about $\Delta m_0 \approx 2\times10^{-4}$ g of the basic stable isotope Mn^{55}, which constituted $N_0=1.8\times10^{18}$ atoms of Mn^{55}. The total amount of admixed iron (as a chemical element) did not exceed $\Delta m_1 \approx 10^{-8}$ g.

Before preparation of the nutrient water-salt medium, all dry ingredients (reagents) were investigated using the Mossbauer gamma-spectrometer to control the content of the admixed Fe^{57} isotope.

Similar preliminary studies with the Mossbauer gamma-spectrometer were carried out on the portion of desiccated microbiological culture before its use in the transmutation experiments.

In all cases, the content of the admixed Fe^{57} isotope was below the level of detection (the level of sensitivity) of the Mossbauer gamma-spectrometer, which complied with the required specifications for the reagents used.

A typical series of experiments on nuclear transmutation of isotopes in growing microbiological cultures involved simultaneous growing of separated parts of the same culture in several (usually four) flasks (Fig. 3.1).

he first and the second flask contained basic (constant) ingredients: sugar-salt nutrient medium on the basis of light water (H_2O) both with and without $MnSO_4$. In the third flask, the nutrient medium was prepared from the same basic ingredients on the basis of heavy water (D_2O), but without $MnSO_4$. Accordingly, the fourth dish contained nutrient medium with all the ingredients necessary for the culture's growth as well as $MnSO_4$, required for transmutation, and was prepared on the basis of heavy water (D_2O).

From the list of nutrient media, necessary for growing cultures, it can be seen, that the isotopic composition necessary for achieving transmutation was in only one (optimized by the isotopic and elementary content)—the fourth flask.

The method of cross-combinations of the nutrient media ingredients has allowed excluding a possible influence of the admixed Fe^{57} isotope on the result of these experiments. In particular, if the Fe^{57} isotope was present, in the form of admixture, in light or heavy water, as part of basic salts, and was also part of the flask's glass, or contained in the air, that isotope would be extracted during the culture's growth and detected in all flask (including the experiments in flasks *1–3*) after growing the culture.

If the Fe^{57} isotope were found only in heavy water, it would be detected in the cultures grown in flasks *3* and *4*. If it was present in $MnSO_4$, it would also be detected in the cultures grown in flasks *2* and *4*.

The Mossbauer analysis has shown that Fe^{57} isotope is present in recognizable quantities only in cultures that were grown in flask *4*, i.e. with all the conditions in place necessary for the low-energy synthesis.

Such series of experiments were performed for various cultures, with different growth periods (24, 48, and 72 hours) and growth modes. All cultures were grown in the thermostat with the optimal temperature of 32° C. It was discovered, that growing cultures in the media based on heavy water requires continuous stirring of the medium throughout the whole time of growth.

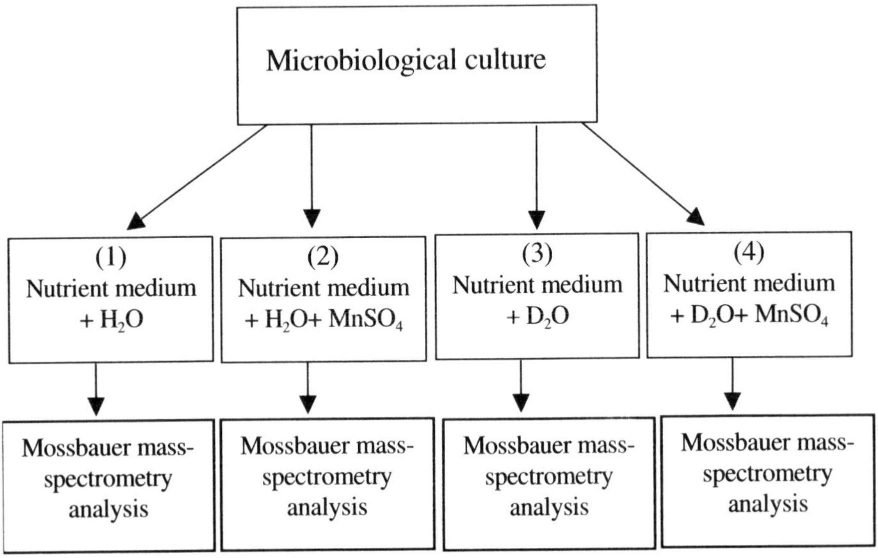

Fig. 3.1. The scheme of "cross-experiments" for studying the effect of nuclear transmutation of isotopes in growing microbiological cultures

Once each of the series was completed, the obtained biological substance was separated from the nutrient medium in a centrifuge, thoroughly rinsed by light distilled water (H_2O), and then dried. Dried substance in the form of fine film was later separated by non-metallic (glass) instruments, grinded to powder, and a portion of it placed in the Mossbauer spectrometer (the rest of the desiccated culture was used for conducting independent tests on a laser fly-through mass-spectrometer).

In order to increase the probability of resonance absorption in the Mossbauer gamma-spectrometer, some of the experiments involved cooling part of the dry culture to the temperature of liquid nitrogen T = 78 K.

Radioactive isotope Co^{57} with activity 30–40 mCi in the chrome matrix was used as a source of the Mossbauer radiation in the gamma-spectrometer. That source had the specter of gamma-radiation in the form of a single gamma-line of natural width with energy equal to 14.4 KeV.

To improve sensitivity of the gamma-spectrometer, a conversion detector of the Mossbauer resonance gamma-radiation was used, which allowed to increase sensitivity of detection of the synthesized Fe^{57} isotope by α=9 times compared to traditional (non-conversion) gamma-detectors. Here, $\alpha$is the coefficient of internal conversion of electrons for the Mossbauer gamma-transition in the Fe^{57} nucleus.

Measurements in each of the 256 channels of the gamma-spectrometer were taken until $N_\infty = 3.5\times10^6$ measurements in the channels of the spectrometer, distanced from the gamma-resonance line, had been made. Such statistic provided the necessary relative precision of measurements, equal to

$$<\Delta N>/<N> \approx 5.10^{-4} = 0{,}05\%.$$

Fig. 3.2a shows the Mossbauer spectrum for the culture *Saccharomyces cerevisiae T-8*, grown for $\tilde{t}$ = 72 hours in the optimal nutrient media, corresponding to flask *4*, i.e. containing the necessary ingredients for achieving transmutation (D_2O and $MnSO_4$).

It can be seen, that in the spectrum there are clear signs indicating the presence of the Mossbauer Fe^{57} isotope. The total mass of the grown culture in each of the experiments was equal 0.3–0.5 g. Spectrum 3.2b corresponded to the same culture *Saccharomyces cerevisiae T-8*—grown in flask *2*—containing the medium based on light water with the presence of $MnSO_4$, and spectrum 3.2c also related to the same culture grown in flask *3* (medium based on heavy water D_2O, without adding $MnSO_4$ solt).

It follows from these spectrums, that the amount of Fe^{57} isotope in cultures grown in non-optimal mediums matches the background level and lies below the level of detection.

The natural width of Mossbauer resonance in Fe^{57} isotope (in units of velocity) is equal Γ= 0.19 mm/sec. In spectrum 3.2a related to the culture grown in optimal conditions, the registered gamma-line was made broader by

$$p_a = \Delta\omega_a/\Gamma \approx 10 \text{ times}$$

in relation to the natural width Γ.

The nature of such broadening is related to interaction of quadruple moments of the nuclei belonging to the Fe^{57} isotope being studied with the gradient of electrostatic field of desiccated and ground to powder culture.

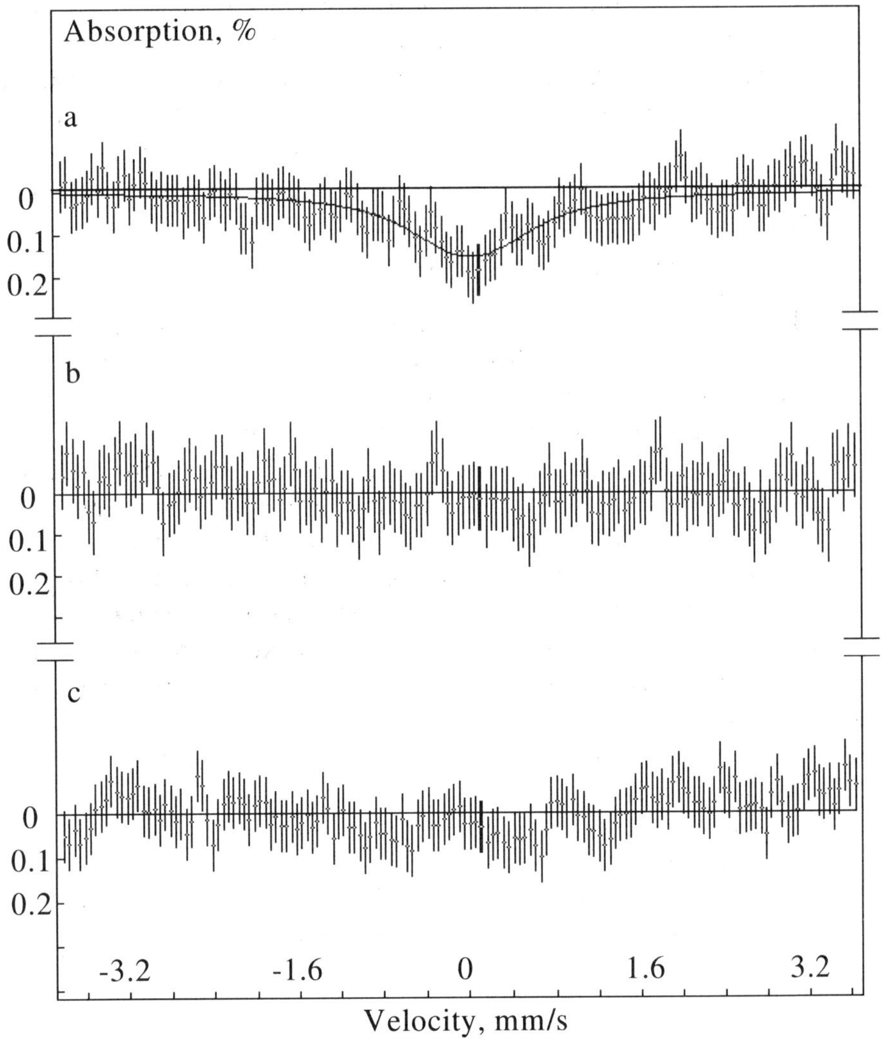

Fig. 3.2. The Mossbauer specter for the grown culture *Saccharomyces cerevisiae T-8*: *a*) in D_2O with Mn^{55}; *b*) in H_2O with Mn^{55}; *c*) in D_2O without Mn^{55}

The total number of nuclei of iron, containing the Mossbauer resonance isotope Fe^{57} with the Lorentz-type broadening was calculated according to the formula

$$N_a = \frac{\eta_a S(1+p_a)}{\sigma_0 f}$$

for a thin Mossbauer absorber.

Here, $\eta_a = \frac{N_\infty - N_0}{N_0}$ - relative depth of the absorption resonance; N_∞ and N_0—the number of counts in the remote and the resonance channels of the gamma-spectrometer, $S \approx 0.63 cm^2$— cross-section of the gamma-spectrometer's window, filled with desiccated experimental culture, $f \approx 0.6$— probability of the Mossbauer effect in the absorbent made of the experimental culture, σ_0—cross-section of resonance absorption in the Fe^{57} isotope.

Final values for the relative depth of absorption resonance, the total number of nuclei of the synthesized Fe^{57} isotope and the mass of these nuclei are equal, correspondingly:

$$\eta_a = (1.9 \pm 0.53).10^{-3},\ N_a = (0.87 \pm 0.24).10^{16},\ m_a = (1 \pm 0.28).10^{-6}\ g.$$

The transmutation coefficient, corresponding to the detected reaction of low-temperature synthesis, is equal to

$$\lambda_a = \frac{N_a}{N_0 t} = (1.9 \pm 0.52).10^{-8}\ s^{-1}$$

(nuclei of the synthesized Fe^{57} isotope, counting for 1 sec and a single nuclei of the Mn^{55} isotope).

Similar results were obtained in the process of growth of other cultures in optimal conditions.

In some cases (for example, for the culture *Deinococcus radiodurans M-1*) the specter structure of the Mossbauer absorption for the synthesized isotope Fe^{57} in the grown culture had the form of a doublet, which reflected the altered spatial structure of a desiccated biological substance.

Fig. 3.3a and 3.3b show experimental specters, representing the results of studies conducted on the Mossbauer gamma-spectrometer with the culture *Deinococcus radiodurans M-1*, grown in different series of experiments in optimal conditions in the media based on heavy water D_2O with the presence of $MnSO_4$.

The total number of synthesized nuclei of the Fe^{57} isotope was in this case determined by the formula

$$N_a = \frac{2\eta_a S(1+p)}{\sigma_0 f}$$

The final values for this microbiological culture for the relative depth of absorption resonance, the total number of nuclei of the synthesized Fe^{57} isotope and the mass of these nuclei are equal:

$$p_a=3.4,\ \eta_a = (1.53\pm0.57).10^{-3},\ N_a = (0.59\pm0.2).10^{16},$$
$$m_a = (0.67\pm0.24).10^{-6}\ g.$$

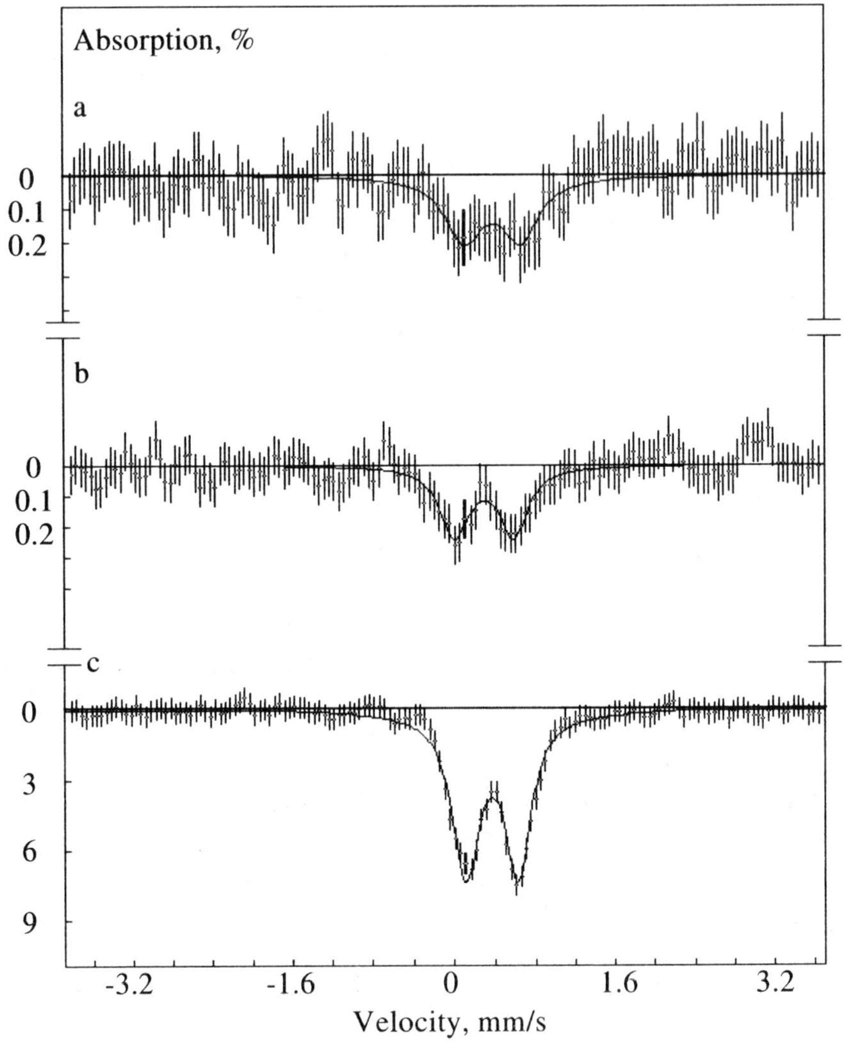

Fig. 3.3. The Mossbauer specter for the grown culture *Deinococcus radiodurans:* a) and b) in D_2O with Mn^{55}; *c)* in D_2O without Mn^{55} and with admixture of small amount of Fe^{57} isotope

The corresponding value of the transmutation coefficient in the growing culture *Deinococcus radiodurans* is equal

$$\lambda_a = (1.3 \pm 0.46).10^{-8}\, s^{-1}$$

(nuclei of the synthesized Fe^{57} isotope, counting for 1 sec and a single nuclei of the Mn^{55} isotope).

In the course of conducting the experiments, a thorough check for artifacts was performed.

One of the hypothetical sources of artifacts was related to potential influence of the nutrient medium and its ingredients on biological assimilability of admixed Fe^{57}. Owing to the lack of preliminary experiments, investigating the specifics of involving various isotopes of iron in the process of culture growth, this problem acquired special attention. For example, the hypothesis stating that a Fe^{57} isotope is assimilated by a growing culture in D_2O and rejected in the media based on H_2O, is correct. Another suggestion may be related to the hypothesis about a different structure of gamma-absorption of Fe^{57} in cultures grown on the basis of D_2O and H_2O (for example, the culture containing H_2O and $MnSO_4$ is distinguished by a multi-component extra thin structure with a corresponding decrease of each of the specter's components to the level, which is undetectable in an experiment, while the culture, containing D_2O such separation is not found).

To verify possibility and grounds for such effects, we performed an investigation of the gamma-spectrum of grown cultures when, during the growth process, a limited controlled small quantity of Fe^{57} isotope was added to the non-optimal nutrient medium (containing H_2O and $MnSO_4$). Fig.3.3c shows the gamma-spectum for this case of stimulated assimilation of Fe^{57} nuclei by the *Deinococcus radiodurans M-1* culture in a non-optimal medium, which is completely analogous to the gamma-spectrum in experiments on transmutation (Fig. 3.2a, 3.2b) in an optimal medium (containing D_2O and $MnSO_4$), proving insignificance of this hypothetical source of artifacts.

Thus, appearance of absorption resonance, corresponding to the Fe^{57} isotope in the transmutation experiment is not correlated with transformation of the super fine structure and spectral transformation of specific components of the resonance gamma-absorption curve, but serves as a reliable evidence of the Fe^{57} isotope synthesis in reaction of transmutation.

In the end, several conclusions can be made.

The experimental data has shown that:

(1) Isotope Fe^{57} is not found in significant quantities in the ingredients of the nutrient medium;
(2) Isotope Fe^{57} is not present in heavy water and cannot be extracted from glass or air during the microbiological culture's growth;
(3) The structure of the gamma-absorption spectrum of Fe^{57} isotope in cultures grown in the media based on light and heavy water is the same.

Therefore, Fe^{57} isotope, detectable after completion of the microbiological cultures' growth in the optimal nutrient medium is not related to artifacts and is generated in the low-temperature reaction of nuclear transmutation

$$Mn^{54} + d^2 = Fe^{57}$$

in the process of these cultures' growth.

Finally, we shall examine thermodynamic consequences of the given reaction.

Since the synthesis reaction $Mn^{54} + d^2 = Fe^{57}$ is energetically efficient and should run with energy emission, the average power $\langle P_a \rangle$, which could be generated in the dish with the optimal nutrient medium, can be calculated from the difference of mass defects of the nuclei participating in the reaction

$$\langle P_a \rangle \approx N_a\,[M(Mn^{55}) + M(d^2) - M(Fe^{57})]\,c^2 / \Delta t$$

In this expression $M(Mn^{55})$, $M(d^2)$ and $M(Fe^{57})$—are masses of isotopes Mn^{55}, d^2 and Fe^{57}.

The difference of mass defects of the isotopes participating in the reaction of transmutation is equal

$$\Delta Mc^2 = [M(Mn^{55}) + M(d^2) - M(Fe^{57})]c^2 = 15{,}605 \text{ MeV}.$$

Using the experimentally obtained values, $N_a \approx (0.9 \div 0.6).10^{16}$, $\Delta t \approx 3.10^5$ sec, we calculated that the average power, generated during the process of transmutation in a flask with volume $V = 10$ cm^3, does not exceed the value of

$$\langle P_a \rangle \approx 0.075 \div 0.04 \text{ W}.$$

Leaving aside the special question of distribution of this energy among products of the reaction, we note, that this power—even if fully conversed into heat—could not significantly alter the thermal regime of culture's growth (especially provided that the growing process was performed in an automatic thermostat).

3.2.2 *Study of controlled transmutation of light and intermediate isotopes in a growing microbiological culture using the laser time-of-flight mass-spectrometer*

As we said before, providing maximal reliability of experimental results is one of the principal requirements to experiments on transmutation of isotopes. This goal can only be achieved by the use of several independent methods of investigation of the same object.

We conducted an alternative study of mass specters of the cultures, which were grown in nutrient media, optimized for the reaction of transmutation

$$Mn^{54} + d^2 = Fe^{57},$$

and then analyzed in the Mossbauer gamma-spectrometer.

Research was conducted on a laser time-of-f light mass-spectrometer, designed for making local microanalysis of solid materials with a single laser flash, focused on the object of investigation. Proceeding from the same requirements of reliability of experimental results, let's consider in detail the operational features of the device being used in the experiments.

The principal working scheme of the device is shown on Fig. 3.4. A single ruby-based impulse laser *1* was used for generation of low-temperature plasma (emission wave length λ = 0.63 micron) with impulse energy W = 0.2 J and duration τ = 40 nsec.

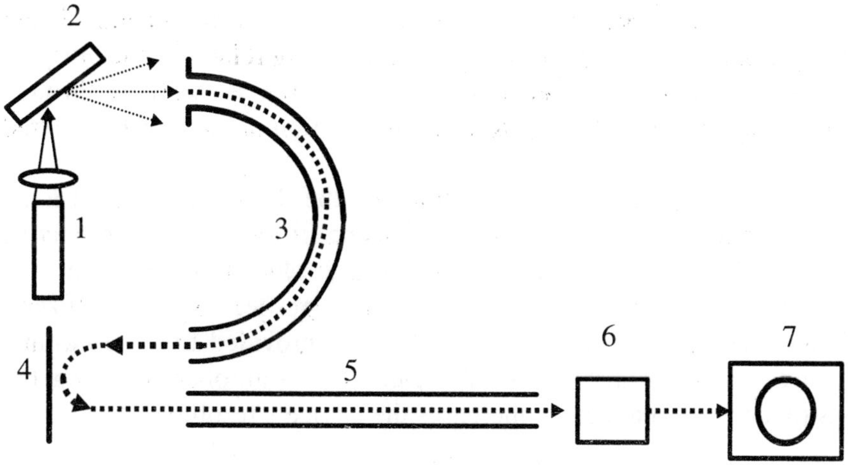

Fig. 3.4. Scheme of the time-of-flight laser mass-spectrometer

Diameter of the laser's focal point D = 100 microns provided generation of laser plasma with the temperature kT≈ 100 eV on the surface of the investigated object *2*. This energy is optimal, based on mutually exclusive requirements—increasing the temperature for a reliable generation of pure atomic (and not molecular) plasma and its lowering for a guaranteed obtaining of a single degree of ionization. As a result of making each laser impulse there was an emission of atoms belonging to the sample on the surface of the studied object with the diameter 100 microns and depth of about 1 micron.

Due to a relatively high (in the scale of values characteristic for molecular links) plasma temperature, all molecular links of the investigated material were completely severed and generation of single (about 95%) and double (about 4%) ions took place. These ions freely dispersed away from the focal point of the laser.

Division of clotted ions of different isotopes was taking place after the energetic discriminator 3, which picked only ions with desired energy W_0≈ kT≈ 100 eV out of the entire thermal specter. Detection of clotted ions after their flythrough selection was due to the difference of velocities

$$v_i = \sqrt{2W_0 / m_i}$$

with the constant energy W_0 in the analyzer's section *5*. There was breaking up of clots of various elements and isotopes. The best conditions for observation correspond to isotopic components with low concentration.

The method of time-of-flight mass-spectrometry allows performing simultaneous and mutually independent analysis of the chemical and isotopic content of the samples through measuring relative concentration of separate components (as in the table below), as well as through investigating the period of space-time modulation of each of the clotted ions.

To improve analytical capabilities of the device, an optical-ionic system was used allowing 4 - phase focusing of ions in a flat reflective field. Placing this system relative to the exit slot of the analyzer and selection of the discharge angle for ionic packages leaving the analyzer *5* with the size of the drift area about 100 cm provided their space-time focusing near the entry slot of the electronic multiplier *6*. Using the focusing optical-ionic device increases the device's resolution to

$$R = m_1/(m_2 - m_1) = 600$$

with the present relative sensitivity of detected elements.

Processed electronic signals were sent to the double-beam oscillograph *7* with memory, which were later recorded on photo film or saved in a computer for further processing.

Fig. 3.5 provides the results of measurements of mass-spectrums in the region of masses, including mass numbers from A = 54 to A = 57 (where the isotopes participating in the reaction are present).

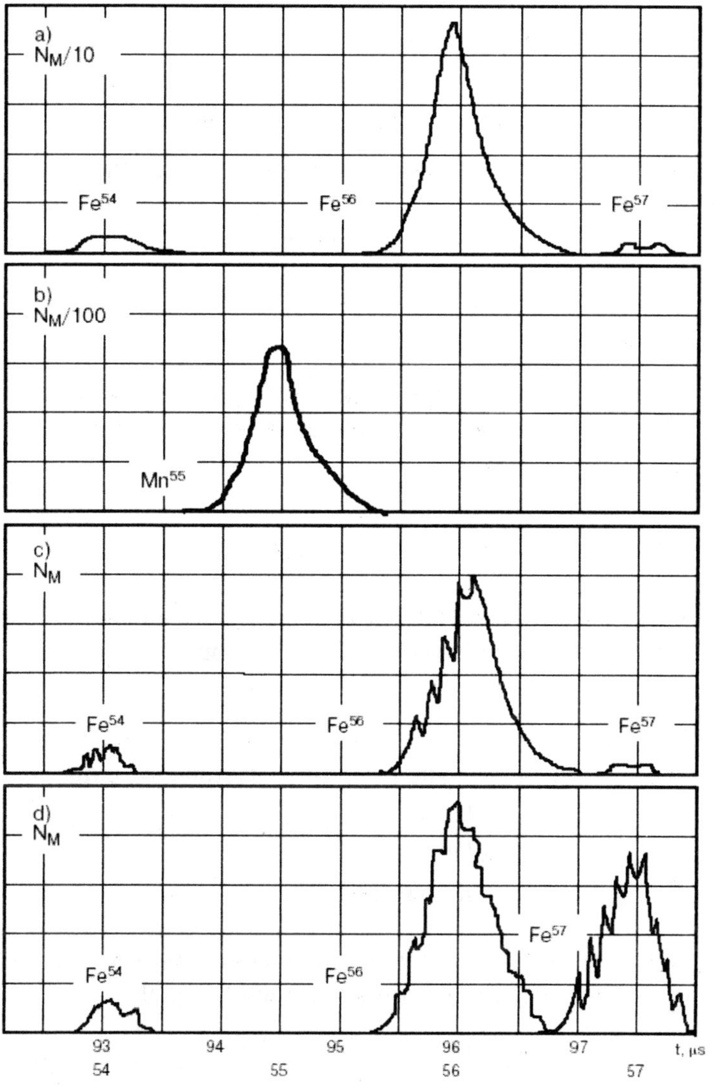

Fig. 3.5 Study of the synthesis process of the Mossbauer isotope Fe^{57} by time-of-flight laser mass-spectrometer.

Fig. 3.5a and 3.5b represent reference spectrums of natural multi-isotope iron and natural single-isotope manganese. Fig. 3.5c gives the results of a control experiment and shows the mass spectrum of the culture *Saccharomyces cerevisiae* grown in the nutrient medium on the basis of heavy water D_2O, but without adding Mn^{55} isotope (without adding $MnSO_4$). On Fig. 3.5d there are the mass spectrum of the same culture grown in the nutrient medium based on heavy water D_2O with added Mn^{55} isotope ($MnSO_4$ added).

This spectrum is, in fact, the sum of the synthesized isotope Fe^{57} and admixed isotopes of natural iron (Fe^{54}, Fe^{55} and Fe^{57}).

It follows from the measurement results, that the concentration ratio of isotopes Fe^{57} and Fe^{56} matches its natural value, both in the control experiment (Fig. 3.4c) and in natural iron (Fig. 3.4a)

$$\frac{\eta(Fe^{57})}{\eta(Fe^{56})} \approx 0.02$$

In contrast to that, the ratio of the same isotopes' concentrations in a culture grown in the optimal nutrient medium is above the natural level by 35–45 times and equals

$$\frac{\eta(Fe^{57})}{\eta(Fe^{56})} \approx 0.7 - 0.9 .$$

This increase corresponds to a very significant rise of the mass content of the synthesized isotope Fe^{57} in relation to the admixed isotope Fe^{56}.

Taking this change into account, it follows, that the total mass and the number of atoms of the synthesized Fe^{57} isotope are equal

$$m(Fe^{57}) = \frac{\eta(Fe^{57})}{\eta(Fe^{56})} m(Fe^{56}) \approx (0.8 - 1.0).10^{-6}\, g,$$

$$N(Fe^{57}) \approx (0.6 - 0.8).10^{16}$$

The corresponding coefficient of transmutation (obtained by using the method of time-of-flight mass-spectrometry) equals

$$\lambda^{(Mass-spectr)} = \frac{N(Fe^{57})}{N(Mn^{55})\Delta t} \approx \frac{\eta(Fe^{57})}{\eta(Fe^{56})} \frac{N(Fe^{56})}{N(Mn^{55})\Delta t} \approx (1.0 - 1.3).10^{-8}$$

(nuclei of the synthesized isotope Fe^{57} counting for 1 sec and the single nuclei of the Mn^{55} isotope).

Therefore, it can be observed, that the coefficients of transmutation obtained by using the methods of Mossbauer gamma-spectrometry and time-of-flight mass-spectrometry are approximately equal.

3.3 Monitoring and mass-spectrometry analysis of controlled transmutation of isotopes with middlerange atomic numbers in a growing culture

3.3.1. *Experimental methodology and procedures*

Experiments on monitoring and research of the phenomenon of nuclear low-temperature transmutation of isotopes with middle-ranged atomic numbers were based on the expected synthesis reaction

$$Na^{23} + P^{31} = Fe^{54}$$

in a growing microbiological culture placed in an optimized nutrient medium, with deficit of iron, but additionally containing controlled quantities of single-isotope chemical elements Na^{23} and P^{31}.

This reaction is characterized by a positive energy emission

$$\Delta E = [M(Na^{23}) + M(P^{31}) - M(Fe^{54})]c^2 = 22{,}4 \text{ MeV}$$

and, therefore, can run with any energy of interacting isotopes (including close to zero energy of relative motion). The main problem here is, naturally, breaking the Coulomb's barrier. It will be examined in another chapter. Research was conducted on the bacterial culture *Bacillus subtilis*.

The result was a rare isotope of iron Fe^{54}, concentration of which in natural iron is small and equal to $\eta = 5.8\%$. An obvious advantage of this reaction (as the synthesis reaction of the rare Mossbauer isotope Fe^{57}) is stability of its final product, its low concentration in natural iron and ability to accumulate in the volume of a culture during its growth. It can be easily detected with the help of mass-spectrometry.

The following methodology of conducting experiments was used.

A prepared culture, after being treated in a centrifuge, rinsed with water and preliminary bred, was placed in pans with sugar-salt nutrient medium containing salts of Mg, Ca, K, Na, sucrose and 100 ml of distillated light water H_2O. The culture was grown in a thermostat at the temperature of 32°C, optimal for growth. The make-up of the nutrient medium is presented in Table 3.2.

Table 3.2

Nutrient medium ingradients	*Concentration in the medium (%)*	*Admixtured iron (no more than), %*	*Admixtured iron (no more than), g*
Sucrose	2	10^{-4}	2.10^{-4}
$MgSO_4$	0,05	2.10^{-4}	10^{-5}
$CaCO_3$	0,2	$1,5.10^{-4}$	3.10^{-5}
KCl	0,05	3.10^{-4}	$1,5.10^{-5}$
$NaNO_3$	0,5	2.10^{-4}	10^{-4}
K_2HPO_4 (in experiments on transmutation)	0,2	5.10^{-4}	10^{-4}
Light distilled water H_2O	100 (100 ml)	10^{-7}	10^{-5}

The main idea of the experiments was growing cultures in several alternative nutrient mediums, only one of which containing both Na^{23} and P^{31} isotopes, was required for a synthesis reaction. For this purpose, in the experiment on transmutation, to the nutrient medium containing, among other substances, $NaNO_3$, was added K_2HPO_4.

The total mass and the number of atoms of isotope Na^{23} in $NaNO_3$ were equal

$$m(Na^{23}) \approx 0.13 \text{ g}, N(Na^{23}) \approx 2,8.10^{21}.$$

The amount of admixed iron in this salt did not exceed a small value. $\Delta m_{Fe} \approx 10^{-4}$ g, and the mass and number of atoms of isotope P^{31} was equal to

$$m(P^{31}) \approx 0.1\text{g}, N(P^{31}) \approx 1,1.10^{21} \text{ atoms}.$$

All dry components of the nutrient medium were preliminary analyzed on a mass-spectrometer. Same investigations were carried out with part of the culture subject to growing. Admixed iron content was found to be small and approximately matched the value indicated in the specification.

A typical series of each of the experiments on nuclear transmutation consisted of independent growth of separated parts of the same culture in 2 flasks with different nutrient media (see Fig. 3.6).

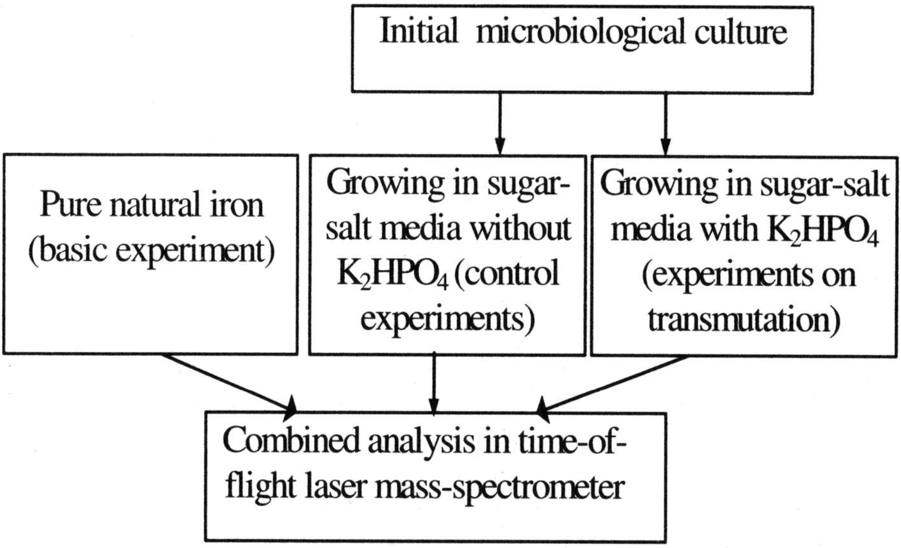

Fig. 3.6 The experimental scheme on transmutation and spectrometry of isotopes in microbiological cultures

The first flask contained water-salt nutrient medium with all ingredients (including $NaNO_3$), but without K_2HPO_4, and the second one—the same media with all the ingredients, but with K_2HPO_4. Duration of growing cultures in different series was 24, 48, or 72 hours.

A typical series of measurements of mass specters consisted of three consecutively administered main operations:

(a) measuring the mass specter of pure natural iron (the basic experiment for obtaining the basic benchmark mass specter of all natural iron isotopes);

(b) research of the mass specter of the reference desiccated culture, grown in the medium with all ingredients (including Na^{23} isotope contained in $NaNO_3$), but without the P^{31} isotope (i.e. without K_2HPO_4);

(c) research of the mass specter of a desiccated culture grown in the medium with all ingredients with the presence of Na^{23} and P^{31} isotopes.

Prior to the making measurements and creating vacuum, the studied objects were placed in the same vacuum camera of a laser mass-spectrometer. All samples were placed in open separated micro cavities, made on the surface of a high-purity lead plate, which were open from the side of laser impulse exposure, with a diameter greater than the cross section of the focal point of the laser.

The adjusting system of the mass-spectrometer could move this plate inside the vacuum chamber and place each of the micro cavities on the point coinciding with the focal point of laser impulse. After each sample has been subjected to laser impulse impact, a cloud of dispersing laser plasma with a temperature of about 100 eV was formed. After achieving extraction of monoenergetic ions from plasma (using an energetic discriminator with narrow energetic window) and subsequent fly-through selection, plasma ions were separated according to their mass. After detection, processed electronic signals were sent to a double-beam oscillograph with memorizing capacity with subsequent storing on photo film or inside a data storing and processing system of a computer.

First, an analysis of the entire mass specter was conducted, and then separate isotopes were investigated in more detail.

Reliability and presentation of recording measurement results were given special attention. The most clearness was achieved with photographing mass specters of isotopes on the double-beam oscillograph with memory. Registration of isotopes' distribution was accomplished in the form of a sequence of paired series of measurements.

Each one of those paired series initially included obtaining the basic (benchmark) specter for natural iron isotopes, which was formed by influencing micro cavities containing iron by laser impulse. The basic specter was then recorded in the form of a synchronized with the laser impulse upper beam sweep on the screen of the double-beam oscillograph. After that, the micro cavity with a sample of an investigated culture was moved to the area of laser focal point and a specter of isotopes of that culture was obtained. That specter was recorded on the same screen of the double-beam oscillograph in the form of the lower beam sweep, which was fully synchronized with the upper beam sweep representing the basic specter of natural iron. After that, obtained specters were photographed from the screen of the oscillograph.

The same specters were independently registered on a computer.

3.3.2. *Results and discussion*

Computer processed results of the experiments are presented on Fig. 3.7 and photographs on Fig. 3.8.

The photographs (Fig. 3.8) are made from the screen of a multi-beam oscillograph with memory and show the mass specter of the basic and control samples, as well as the samples, which were subjected to the process of transmutation. The upper beam sweep (upper curve) on all figures corresponds to the specter for natural iron (the basic experiment), and the lower beam sweep—to the specter of a studied culture (control or transmutation experiments).

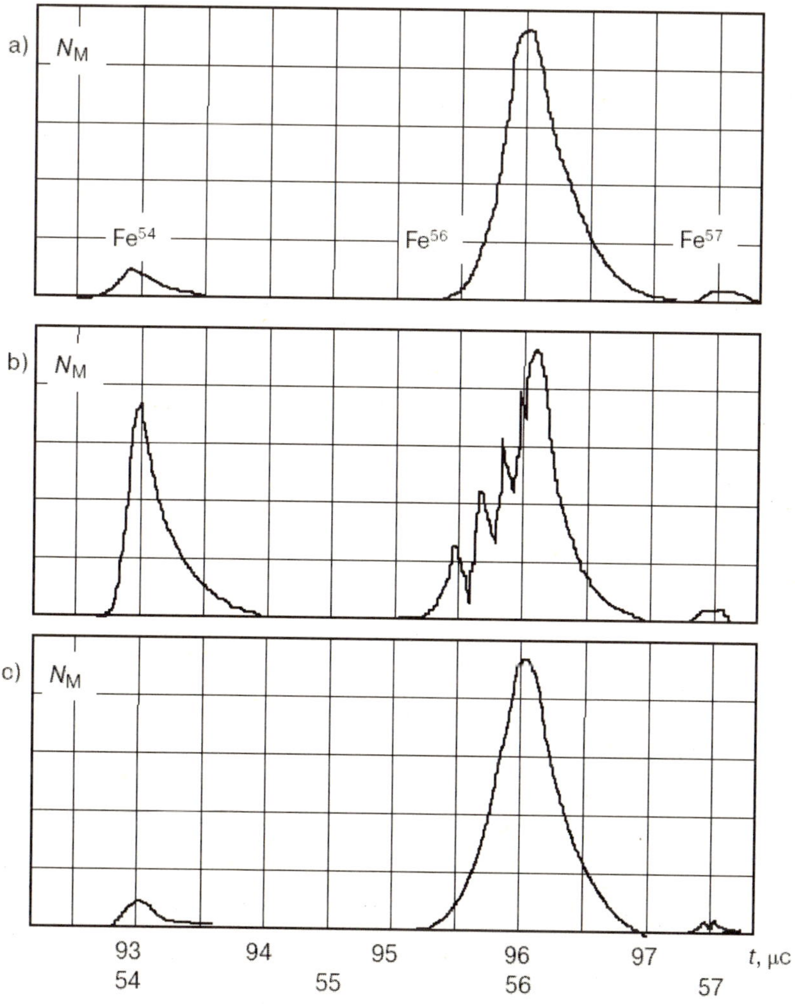

Fig. 3.7 Typical mass specter series for isotopes of iron: a) the basic (benchmark) experiment for pure natural iron; b) experiment on transmutation (the culture is grown in a medium in the presence of P^{31} and Na^{23} isotopes); c) controlling experiment (the culture is grown in a medium without P^{31} isotope). t—time of flight of ions of isotopes from target to detector, 54...57 —atomic number of isotope.

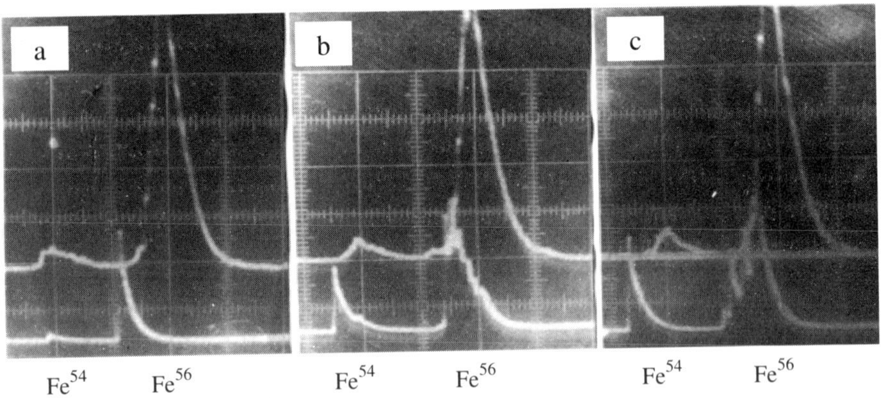

Fig.3.8 Photographs from the screen of the oscillograph, representing the mass specter in the area of isotopes of iron. The upper graphs show the basic (benchmark) experiment for pure natural iron; the lower graphs show the mass specter of grown microbiological culture. a) controlling experiment (culture grown in a medium without isotope P^{31}), b) and c) — transmutation experiments (culture grown in a medium in the presence of isotopes P^{31} and Na^{23}).

It can be seen from the photographs, that in the specters of each of the studied cultures there is a minor shift of the ion current peak corresponding to different isotopes to the left (i.e. towards lower masses), relative to the position of corresponding natural iron isotopes in each specter. This shift is related to the specifics of the time-of-flight spectrometer, determined by special features of movement and interaction (a slight repulsion) of ionic clots, corresponding to different elements in the process of their separation and consequent movement. The influence of the plasma package of heavy ions of lead, which, repulsing the ions of lighter elements and isotopes, slightly shift them on their trajectory, causing their earlier detection, is mainly responsible for such additional shifting of ion clots.

Evident from the photographs and diagrams is a very large increase of absolute and relative concentration of rare isotope Fe^{54} for the culture grown in the optimal nutrient medium. This increase corresponded to deviation of concentration ratio of isotopes from the natural value

$$\frac{\eta(Fe^{54})}{\eta(Fe^{56})} \approx 0.06$$

(for natural iron in the basic experiment and rare admixed natural iron in the controlling experiment) to the value.

$$\frac{\eta(Fe^{54})}{\eta(Fe^{56})} \approx 0.2 \quad 0.25$$

(for the experiment on transmutation in the optimal medium).

Recurrence of successful results in these experiments was close to 100%.

Transmutation coefficients, averaged by several series of measurements, in the reaction $Na^{23} + P^{31} = Fe^{54}$ equal

$$\lambda_P = \frac{N(Fe^{54})}{N(P^{31})\Delta t} = \frac{\eta(Fe^{54})}{\eta(Fe^{56})} \frac{N(Fe^{56})}{N(P^{31})\Delta t} \approx (3-6).10^{-10}\,\text{sec}^{-1}$$

(nuclei of synthesized isotope Fe^{54} counting for 1 sec and the single nuclei of P^{31} isotope),

$$\lambda_{Na} = \frac{N(Fe^{54})}{N(Na^{23})\Delta t} = \frac{\eta(Fe^{54})}{\eta(Fe^{56})} \frac{N(Fe^{56})}{N(Na^{23})\Delta t} \approx (1-2).10^{-10}\,\text{sec}^{-1}$$

(nuclei of synthesized isotope Fe^{54} counting for 1 sec and the single nuclei of Na^{23} isotope).

For more reliable measurements, studies of mass specters of all desiccated (initial) components of the sugar-salt nutrient medium were conducted.

The study of the results of direct impact of laser impulse in mass-spectrometer on main (with regards to participating in the synthesis reaction) components of the nutrient medium—$NaNO_3$ and K_2HPO_4—was also important. Measurements have shown, that the concentration ratio of isotopes $\eta(Fe^{54})/\eta(Fe^{56}) \approx 0.06$ remains unchanged, both in the case of laser irradiation of separated salts $NaNO_3$ and K_2HPO_4 and irradiation of thoroughly prepared fine-dyspersated mixture of these salts.

The latter result is particularly important. It allows to draw a conclusion, that formation of isotope Fe^{54} and changing the ratio of isotopes Fe^{54} and Fe^{56}, associated with this process, is actually observed in the process of culture growth in the optimized medium in the presence of $NaNO_3$ and K_2HPO_4, i.e. it represents the result of a low-temperature nuclear synthesis $Na^{23} + P^{31} = Fe^{54}$ and is not related to the thermal effect of laser impulse on the irradiated sample in mass-spectrometer. It proves, that the observed reaction is not the result of "thermal synthesis" with temperature kT≈ 100 eV in the focal point of laser impulse, but can run at the temperature of 32° C in a growing microbiological culture.

3.4 Observing transmutation of heavy isotopes and studying methods of controlling transmutation processes in growing microbiological cultures

3.4.1 *Catalytic effect of cesium on effectiveness of nuclear transformation of medium and heavy isotopes in a growing biological culture*

The purpose of this research was studying the process of transmutation of isotopes at low energy of their relative movement and in the presence of heavy active (with regards to biochemical reactions) elements. First of all, cesium is considered one of such elements, being the biochemical analogue of potassium.

It is well known, that cesium is not one of the vital microelements.

On the other hand, it is also well known, that potassium and cesium are biochemically similar and cesium can substitute potassium in many chemical and biochemical reactions.

The question of chemical analogues and "biochemical substitutes" will be discussed in greater detail in Chapter 5.

Our experiments were aimed at optimization of the earlier investigated reaction of transmutation $Na^{23} + P^{31} = Fe^{54}$ in a growing culture placed in nutrient medium with an additional influence of cesium salts on the process of growth. Reactions were conducted on the basis of the *Saccharomyces cerevisiae* culture.

Moreover, these experiments also provided an opportunity to investigate the simultaneously occurring reaction of transmutation $Cs^{133} + p^{1} = Ba^{134}$ of cesium isotope with formation of a rare stable isotope Ba^{134}. Effectively, this is the first successful research of biological transmutation of heavy isotopes.

The experimental scheme was traditional for microbiological cultures and experiments were conducted at 32°C.

In the control experiment, the previously generated culture—after rinsing and pre-growing—was placed in a dish with sugar-salt nutrient medium containing salts of Mg, Ca, K or Cs, Na, glucose and 100 ml of distillated light water H_2O (Table 3.4).

In the base experiment on nuclear transmutation of isotopes Na^{23} and P^{31} (comprising the composition of the components of the nutrient medium 1.1 and 1.2), Na_2HPO_4, containing isotope P^{31}, was added to the nutrient medium.

In the experiment on catalytic influence and induction (stimulation) of nuclear transmutation, which is described by series 2.1 and 2.2, instead of basic salt KCl, CsCl was added to the nutrient medium. Those two salts are chemically similar and ions Cs can replace ions K during the growth of a microbiological culture.

Table 3.4

Permanent components of nutrient medium	*Concentration in nutrient medium (%)*
Glucose	2
$NaNO_3$	0.5
$MgSO_4$	0.05
Distillated light water H_2O	100
Varying components of nutrient medium in different experiments	
1 KCl	0.05
1.1. Na_2HPO_4	0.1
1.2. Na_2HPO_4 Tryptone	0.1 0.5
2 CsCl	0,05
2.1. Na_2HPO_4	0.1
2.2. Na_2HPO_4 Tryptone	0.1 0.5

Obtained full spectrum of isotopes and elements in a grown culture is presented on Fig. 3.9 and in table 3.5. On Fig. 3.10 there is a magnified fragment of the general mass specter of isotopes, which includes the area of localization of isotopes of iron.

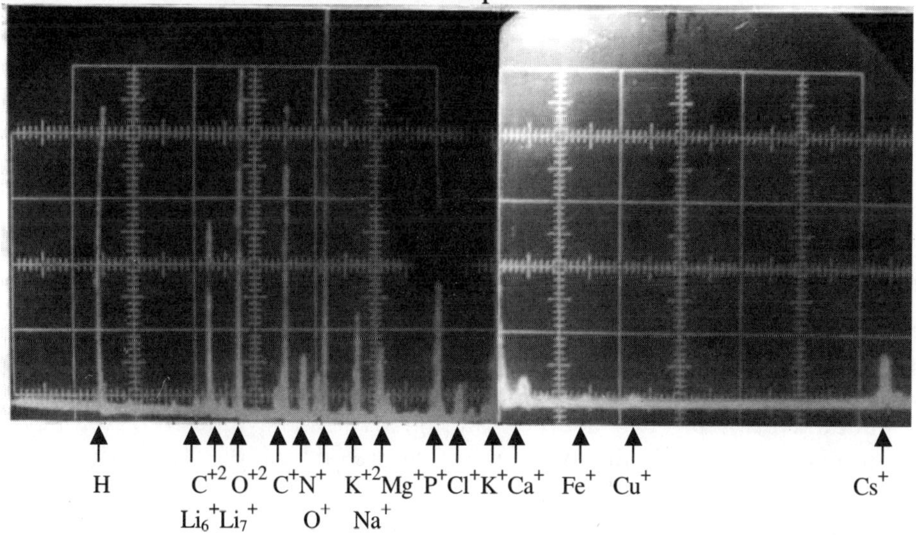

Fig. 3.9 General mass spectrum of isotopes and chemical elements of grown culture Saccharomyces cerevisiae

Each of the mass spectrums was obtained in a time-of-flight spectrometer by a single laser impulse applied to a researched culture.

Table 3.5

Element	*Experiment*			
	1.1	*1.2*	*2.1*	*2.2*
H	0.53	1.0	0.67	0.69
C	23.4	32.1	19.3	18.1
N	3.60	7.90	2.30	6.0
O	21.0	21.3	18.3	25.0
Na	4.80	3.30	5.80	4.40
Mg	0.17	0.40	0.36	0.25
Al	0.30	0.26	0.38	0.04
Si	0.42	0.13	0.67	0.33
P	6.70	11.9	3.47	11.0
S	3.83	7.90	2.22	5.50
Cl	12.6	2.68	5.80	8.80
K	18.6	10.8	13.5	17.0
Ca	0.77	0.62	-	-
Fe	2.5	0.66	2.40	0.27
F	0.03	-	-	0.03
Ti	0.06	-	0.40	0.05
Cu	0.11	-	0.04	0.11
Zn	0.11	-	-	0.11
Cr	0.30	-	0.04	-
Cs	-	-	22.4	2.30
Ba	-	-	1.90	<0.01
Mn	-	0.06	-	0.02

On Fig. 3.10 there is a magnified fragment of the general mass spectrum of isotopes, which includes the area of localization of isotopes of iron.

The mass spectrum of isotopes of natural iron (upper curves on Fig. 3.10a, 3.10b, 3.10c) provides an opportunity to perform precise identification and quantitative interpretation of obtained isotopes (admixed isotopes Fe^{56} and Fe^{54} on Fig. 3.10c and synthesized isotope of iron Fe^{54} obtained in the experiments described by Fig. 3.10a, 3.10b).

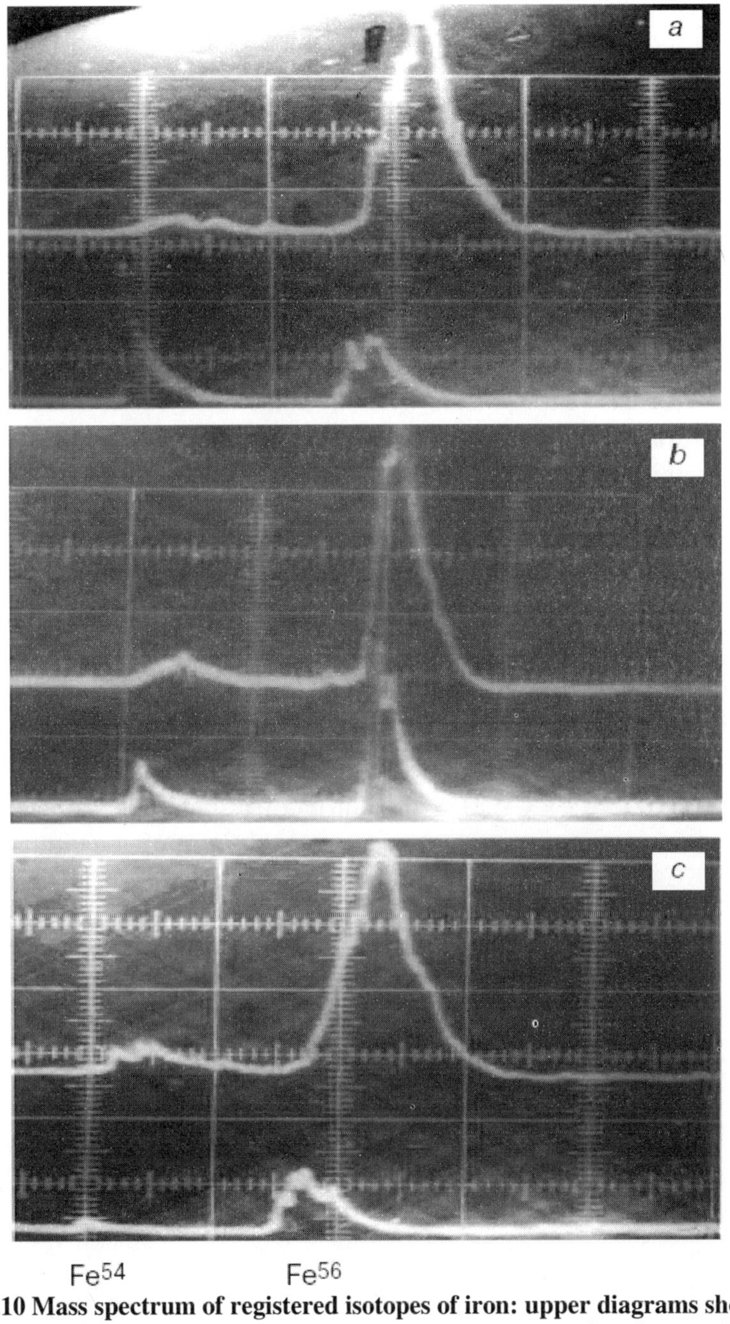

Fig. 3.10 Mass spectrum of registered isotopes of iron: upper diagrams show the basic experiment with natural iron. a) induced transmutation (in a medium containing isotopes Na^{23} and Cs^{133}); b) spontaneous transmutation (in a medium containing isotopes Na^{23} and K^{39}); c) control experiment in a medium without isotope P^{31}. Each of the mass spectrum was obtained in a time-of-flight spectrometer by a single laser impulse applied to a researched culture.

The mass-spectrometry results for the cultures, grown in different nutrient mediums, are the following:

(1) Typically isotopic specter for cultures grown in the absence of phosphorus salt in the nutrient medium (diagrams 1 and 2) is shown on Fig. 4.10c. It can be seen, that, as expected, the ratio of concentrations of isotopes in grown culture. $\eta(Fe^{54})/\eta(Fe^{56}) \approx 0.06$ was the same as in natural iron, while iron itself was of extrinsic origin.

(2) On diagrams 1.1 or 1.2, which corresponded to the absence of cesium salt CsCl in the nutrient medium, experimental results for base experiments were the same as in our earlier works on transmutation on the basis of the reaction $Na^{23} + P^{31} = Fe^{54}$ In this case, in the mass spectrum of grown culture, all chemical elements contained in the nutrient medium (including potassium) were present, but cesium was missing. Part of the mass spectrum, corresponding to the area of localization of isotopes of iron, is presented on Fig. 3.10b. For this case, the ratio of concentrations of isotopes Fe^{54} and Fe^{56} is equal. $\eta(Fe^{54})/\eta(Fe^{56}) \approx 0.2 - 0.25$.

(3) In diagrams 2.1 or 2.2, which correspond to the case when potassium is substituted by chemically similar cesium (KCl is replaced by CsCl) in the nutrient medium, a sharp increase of cesium concentration is registered in grown culture. Since CsCl is fully soluble in nutrient medium based on regular (light) water, including cesium in the makeup of grown culture takes place on the atomic/molecular level, but not in large insoluble clusters. This is an expected and rather trivial result.

The influence of such a replacement on the probability and final effectiveness of sodium and phosphorus transmutation into iron, however, was both not trivial and unexpected.

It follows from Fig. 3.10a, that the presence of cesium in the nutrient medium stimulates the process of transmutation and increases the ratio of isotopic concentration by 1.5–2 times and more up to $\eta(Fe^{54})/\eta(Fe^{56}) \approx 1.0 - 1.5$.

The corresponding transmutation coefficient for such a controlled (stimulated) process is

$$\lambda_{Fe} = \frac{N(Fe^{54})}{N(P^{31} + Na^{23})\Delta t} \approx 10^{-9}\ \text{sec}^{-1}$$

(created nuclei of isotope Fe^{54} per second counting for one pair of nuclei of isotopes P^{31} and Na^{23}).

Control measurements of dry ingredients of nutrient medium have demonstrated, that the curves, corresponding to isotopes of iron with small amplitudes, observed in the general specter, are characterized by the same isotopic ratio $\eta(Fe^{54})/\eta(Fe^{56}) \approx 0.06$ as natural iron. This confirms, that an "extra" amount of isotope Fe^{54} could have been created only in the process of transmutation.

The question about stimulating mechanism of cesium in the synthesis of iron remains open. Cesium does not participate in the paired reaction of iron synthesis as a required element (phosphorus and potassium suffice for that purpose), but it has a strong indirect influence on that process. It is also impossible to explain the fact of synthesis's optimization by the simple proposition, that cesium directly stimulates culture's growth—there is no acceleration of the microbiological culture's growth when potassium is substituted by cesium. A possible explanation of such stimulation is that cesium does not simply influences culture's growth, but affects that part of growth dynamics, which (by our proposition) influences brief removal of the Coulomb's barrier in the synthesis process. This issue will be discussed in more detail in Chapter 4.

3.4.2 *Detection of transmutation of heavy isotopes in growing biological cultures*

In the course of conducting experiments, aimed at studying stimulation of transmutation of potassium and phosphorus into iron by cesium, parallel study of reaction of transmutation of cesium

$$Cs^{133} + p^{1} = Ba^{134}$$

producing a rare barium isotope, natural concentration of which among other isotopes does not exceed 2.4%. The precondition for performing that reaction is described in series 2.1 of the ingredients of nutrient medium (Table 3.4). Despite the fact, that such reaction of transmutation constitutes the process of synthesis and causes generation of an isotope, which is heavier than iron, it is considered energetically productive and characterized by emission of energy

$$\Delta E = [M(Cs^{133}) + M(p^{1}) - M(Ba^{134})]c^2 = 8{,}3 \text{ MeV}.$$

On the photograph (Fig. 3.11), made from the screen of a multi-beam oscillograph with memory, the results of mass-spectrometry of a culture, grown in a nutrient medium optimal for running a transmutation reaction of cesium into barium, are presented. Barium was used as a reference material (basic chemical element).

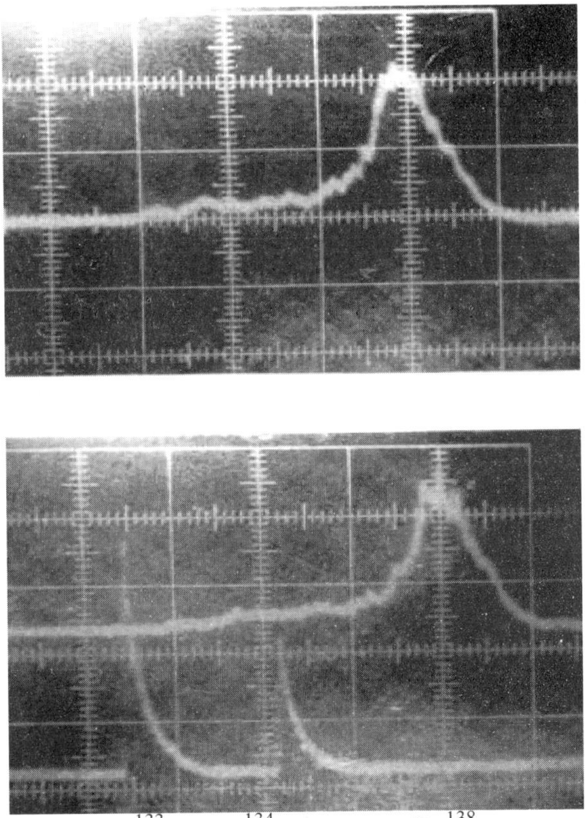

Fig. 3.11 Mass specter of natural barium (the upper graph, in the presence of the base isotope Ba^{138} and more rare isotopes Ba^{134}, Ba^{135}, Ba^{136}, Ba^{137}) and synthesized isotope Ba^{134} in the presence of isotope Cs^{133} in grown culture

On the upper part of Fig. 3.11 there is the mass specter of main stable isotopes of barium. The main maximum corresponds to isotope Ba^{138} (natural content 71.7%). Moreover, there are other maximums in the specter belonging to more rare isotopes of barium—Ba^{134} (natural content 2.4%), Ba^{135} (6.6%), Ba^{136} (7.9%), Ba^{137} (11.2%).

On the upper part of the lower photograph, the mass-specter of barium used as a reference material for identification of isotopes, is presented. On the lower part of the same photograph, there is a fragment of the general mass specter of a desiccated culture grown in a nutrient medium of optimal composition (i.e. in the presence of a single-isotope Cs^{133} and protons p1 contained in water). It can be seen that in the specter there is isotope Cs^{133} contained in the nutrient medium and assimilated by the culture, as well as isotope Ba^{134}, formed as a result of the synthesis re action $Cs^{133} + p^1 = Ba^{134}$.

A slight deformation of specter (a shift to the left of the peak of isotope Cs^{133}, i.e. in the direction of lighter mass, and, correspondingly, the peak of isotope Ba^{134} to the right in the direction of heavier mass) is related to the interaction feature (mutual repulsion), mentioned above, of actual clots of ions of various isotopes in their movement in protracted vacuum channels of a time-of-flight mass-spectrometer.

The approximate equity of the peaks, corresponding to Cs^{133} and Ba^{134}, means a very high effectiveness of cesium transmutation into barium.

Coefficient of transmutation is equal

$\lambda \approx 10^{-8}$ (synthesized nuclei of isotope Ba^{134} counting for 1 sec and one nucleus of isotope Cs^{133}).

3.4.3 *Study of the combined effect of Sr, Cl and S on efficiency of the process of transmutation in biological cultures*

It is well-known, that the accidental (and often unpredictable) nature of low-temperature synthesis is one of the principal characteristics, which distinguishes it from "normal" (high-temperature) synthesis and presents one of the main obstacles hampering its recognition. This circumstance stimulates development of methods of control for such a process.

After unexpected experimental results with substituting potassium for cesium in a medium, where the process of transmutation takes place, it would be logical to investigate the influence of those elements on synthesis (Sr, Cl and S), which, as follows from common micro-biological practice, themselves stimulate or retard growth of cultures.

We conducted a series of investigations dedicated to the study of influence of these elements on the efficiency of reaction of transmutation $Na^{23} + P^{31} = Fe^{54}$ in a growing microbiological culture. None of them participates directly in the reaction of nuclear transformations, but they affect—in different ways—the process of cultures' growth: Cl retards growth most of all, S also has a retarding effect, which is usually much weaker, and the presence of molecular complex (CH_3COO) in the nutrient medium normally causes stimulation of growth in all conditions.

Research was based on bacterial culture *Azotobakter chroococcum*.

Experiments in each of the separate series were distinguished by the following features.

Growth was taking place in media containing traditional and permanent (KCl, KH_2PO_4, $NaNO_3$, pure glucose and distilled light water H_2O), as well as components, which varied according to the task.

In particular, substitution of basic vital microelements (Ca ↔ Sr) through reciprocal substitution of salts $CaCl_2$ and $SrCl_2$ was widely used. From biochemistry it is known, that ions of calcium and strontium are chemically similar and capable of substituting each other.

The second group of replacements, related to studied chemical elements, which were not vital (Cl and S). For higher efficiency and clarity of interpretation, these elements were added to the nutrient medium as components of different soluble salts ($MgCl_2$ ↔ $MgSO_4$ ↔ $Mg(CH_3COO)_3$) belonging to the same chemical element (magnesium), which is a vital microelement necessary for culture's growth.

Composition of all components of the nutrient medium is presented in Table 3.6.

Table 3.6

Nutrient medium components	*Concentration in the medium (%)*
Glucose	2
$NaNO_3$	0.5
KCl	0.05
KH_2PO_4	0,15
$MnSO_4$	0.01
Light distilled water H_2O	100
Controlled (varying) components for transmutation	
1. **$CaCl_2$**	0.5
1.1. $MgSO_4$	0.05
1.2. $MgCl_2$	0.04
1.3. $Mg(CH_3COO)_3$	0.11
2. **$SrCl_2$**	0,5
2.1. $MgSO_4$	0,05
2.2. $MgCl_2$	0,04
2.3. $Mg(CH_3COO)_3$	0,11

During each of the series of experiments researching mass specters, the following measurements were made:

(a) obtaining general mass specter of all elements (from hydrogen to uranium) of grown culture (results of these measurements of concentrations of elements present in the culture are presented in Fig. 3.12 and in Table 3.7);

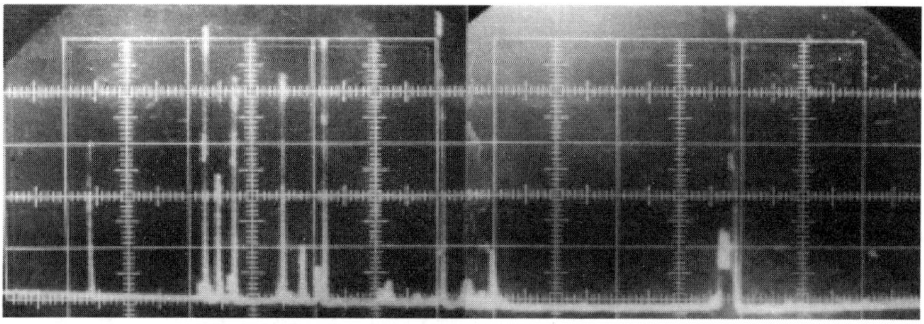

Fig. 3.12. General mass spectrum of a culture grown in a nutrient medium containing strontium

Table 3.7

Element	*Experiment*					
	1.1	*2.1*	*1.2*	*2.2*	*1.3*	*2.3*
H	0.49	0.74	1.44	0.33	0.14	0.58
C	15.0	12.6	31.3	29.3	14.0	11.3
N	5.8	3.8	6.8	1.10	1.63	3.13
O	14.0	11.8	31.3	29.3	13.1	23.4
Na	13.0	5,4	25.8	16.4	1.68	0.46
Mg	0.35	0.5	0.85	0.35	1.0	1.06
Al	0.15	0.16	0.31	0.66	0.15	0.53
P	18.0	17.2	1.1	11.6	27.1	41.2
S	3.0	3.8	-	-	-	-
Cl	7.5	16.5	0.4	1.98	1.0	1.3
K	22.0	18.0	0.15	2.07	39.5	8.65
Ca	0.2	0.8	0.47	0.92	0.34	1.86
Fe	0.1	0.16	<0.01	<0.01	0.13	0.23
Sr	-	5.1	-	4.3	-	5.9
Mo	-	0.26	-	0.38	-	0.18

(b) obtaining base mass-specter of all isotopes of iron (upper curves on Fig. 3.13, 3.14a, 3.14b, 3.14c);

(c) study of mass specter of isotopes in cultures grown in media prepared for transmutation (compositions of nutrient elements 1.1, 1.2, 1.3, 2.1, 2.2, 2.3).

From obtained general mass spectrum for all elements very good assimilability of strontium by a grown microbiological culture can be inferred.

Experimental results are presented in averaged and generalized form on Fig. 3.13.

Photographs from oscillograph screen (Fig. 3.14) demonstrate several typical results for showing dependency of the type of a mass specter from the constitution of the nutrient medium. It follows from this data, that the process of synthesis of isotope Fe^{54} is affected by the presence of Sr and—to a lesser extent—Cl (i.e. transmutation is observed only when salts $MgSO_4$ or $Mg(CH_3COO)_3$ are present, and its stimulation—when salt $Mg(CH_3COO)_3$ is present).

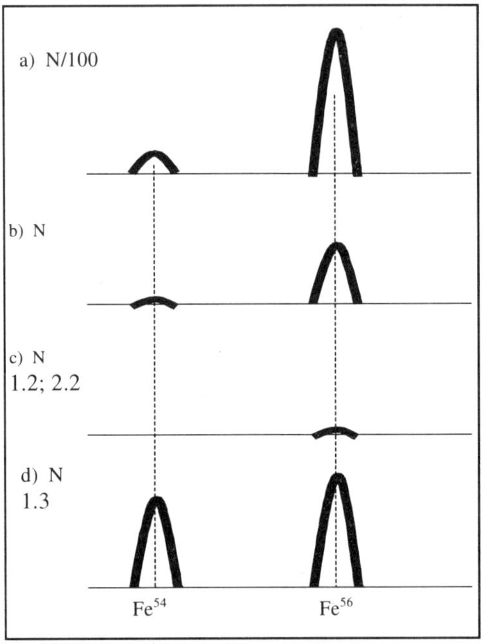

Fig. 3.13. Averaged mass spectrum of registered isotopes of iron. a) specter of natural iron; b), c), d) — spectrum of cultures grown in nutrient media with components 1.1–2.3.

In this case, the rations of concentrations are equal

$$\frac{\eta(Fe^{54})}{\eta(Fe^{56})} \approx 0.2 - 0.4$$

for composition 1.1 and

$$\frac{\eta(Fe^{54})}{\eta(Fe^{56})} \approx 0.5 - 0.6$$

for composition 1.3.

In other cases iron, found in a mass specter, in absent, or found only on the background level (compositions 1.2 and 2.2) or has the isotope structure corresponding to natural iron (compositions 2.1, 2.3), which proves its purely admixed origin.

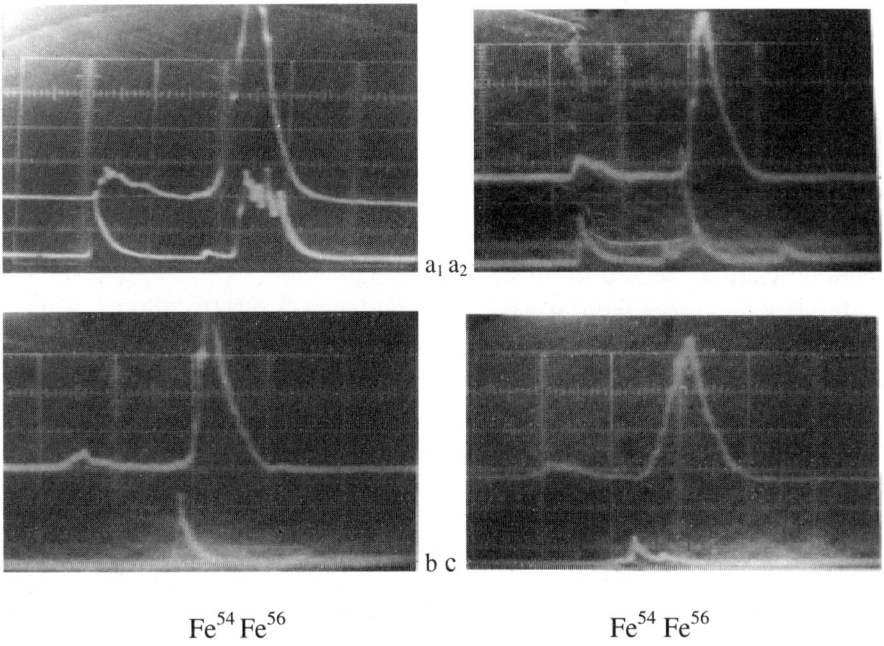

Fig. 3.14. Mass spectrum of registered isotopes of iron: upper curves—spectrum of natural iron, a_1 and a_2—two independent examples of mass spectrums of a culture grown in nutrient medium with composition 1.3 in the presence of isotopes P^{31} and Na^{23}, as well as salt $Mg(CH_3COO)_3$; b—culture grown in nutrient medium with composition 2.1 in the presence of P^{31} and Na^{23} as well as salt $MgCl_2$; c—culture grown in nutrient medium with composition 2.

The general conclusion thus drawn from the data—chemical elements, which partially retard the process of growth of cultures (Sr, Cl and S) also retard the process of transmutation to an even greater degree.

Meanwhile, elements and their salts, which tend to stimulate growth (in our case—$Mg(CH_3COO)_3$), also increase the rate of synthesis and optimize the process of transmutation of isotopes.

3.5. Effective nuclear fusion and transmutation of stable isotopes in microbe syntrophin associations

3.5.1. *The general features of isotopes transmutation in "one-line" cultures and in microbiological associations*

Above we have discussed the process of transmutation of stable isotopes in growing "one-line" (one type, "clean") microbiological cultures like *Escherichia coli* or *Deinicocus Radiodurans*. It was shown that the transmutation process during the growth of such microbiological cultures had taken place but its effectiveness had been low

$\lambda \approx 10^{-8}$ s^{-1} in the case of the reaction with light isotope $Mn^{55}+d^2=Fe^{57}$ and

$\lambda \approx 10^{-10}$ s^{-1} in the reaction with middle range mass isotopes $Na^{23}+P^{31}=Fe^{54}$.

The typical Mossbauer spectrums of "one-line" culture *Saccharomyces cerevisiae* grown in D_2O with the presence of Mn^{55} isotope were presented in Figs 3.2 and 3.3.

The low amplitude of Mossbauer resonance ($\Delta J/J \approx 0.2\%$) in these experiments was the result of low absolute and relative concentration of created Fe^{57} isotope in the culture.

There are two main reasons of low effectiveness of nuclear transmutation in "one-line" microbiological cultures:

1. The relatively low efficiency for creating these reactions is the result of the narrow interval of optimal functional individual characteristics for initiating nuclear activity in any "one-line" type of culture. Each of the "one-line" cultures individually requires a set of specific conditions (temperature, hydrogen ion exponent pH, balanced contents of nutrient medium etc) for achieving optimal metabolic conditions during the complete period of growth. Such conditions are often absent in real experiments.
2. During the growth of a "one-line" culture, we hypothesize that processes involving forms of auto-intoxication of nutrient media by metabolic products take place. This hypothesis is consistent with forms of growth impairment.

In a contrast to these "one-line" cultures investigated micro-biological associations include great number of types of different cultures.

The base of used microbial catalyst-transmutator (MCT) compound is microbe syntrophin associations of thousands different micro-organism kinds that are in the state of complete symbiosis. These micro-organisms appertain to different physiological groups that represent practically the whole variety of the microbe metabolism and relevantly all kinds of microbe accumulation mechanisms. The state of complete symbiosis of the syntrophin associations results on the possibility of maximal adaptation of the micro-organisms' association to any external conditions change.

These cultures are in a state of natural complete symbiosis and grow as a total correlated multisystem. There are a lot of different types of intraspecific and interspecific stimulated and symbiotic connections between different cultures in the volume of syntrophin associations. This correlated microbiological multisystem adequately reacts to modifications of exterior requirements, to composition of nutrient medium and to biochemical properties of a system because of metabolic, growth and transmutation processes. The spectrum of their functional characteristics is very wide. We believe that it should be expected that this would lead to high efficiency for stimulating transmutation processes. This model is presented in symbolic form on Fig. 3.15.

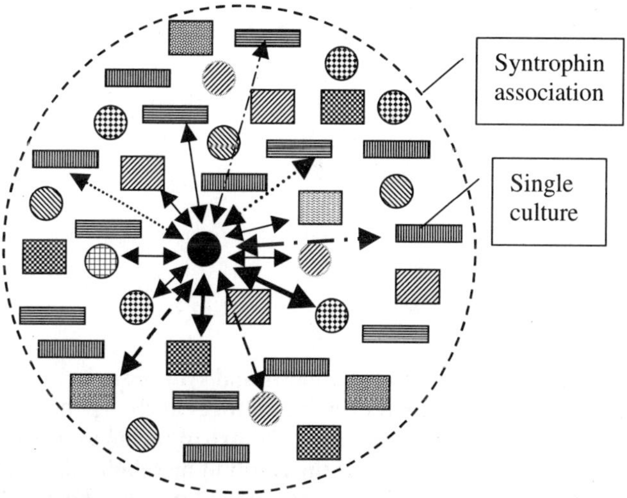

Fig. 3.15. Symbolic scheme of different types of intraspecific and interspecific connections between arbitrary selected culture and different cultures in the volume of syntrophin association. The same connections are related to each culture.

The spectrum of their functional characteristics is very wide. It has been expected that it would lead to high effectiveness of stimulation of the transmutation process. This model is presented in symbolic form in Fig. 3.16.

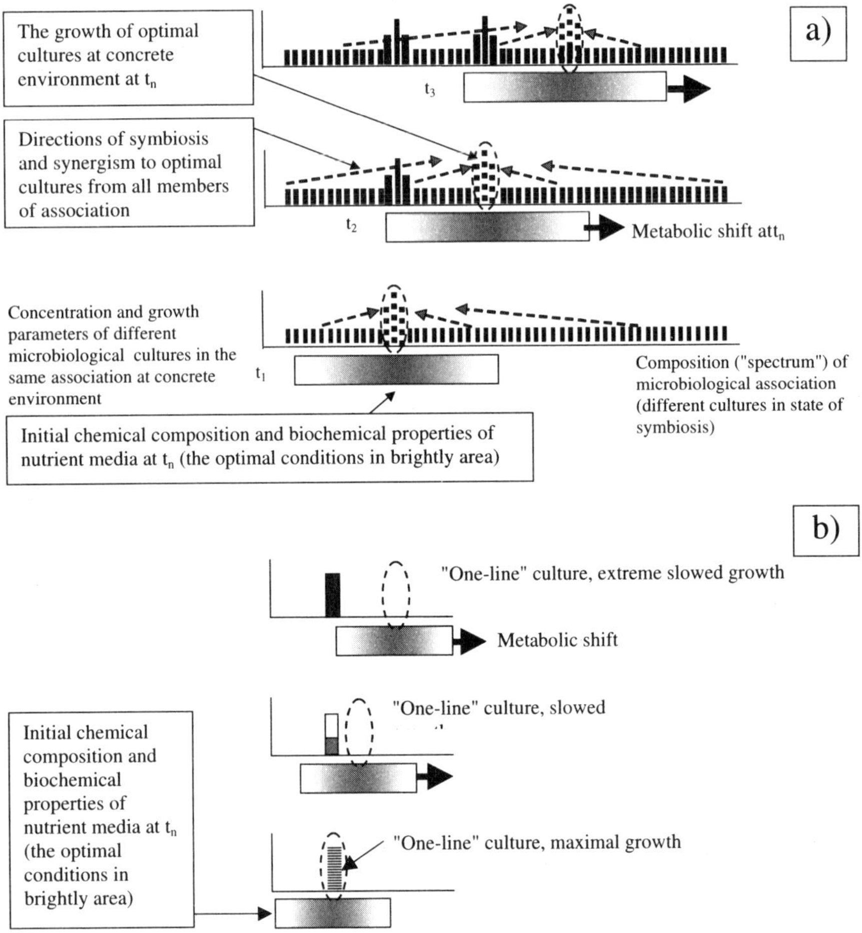

Fig. 3.16. Change the directions of symbiosis and synergism in microbiological association (case a) to optimal growing cultures at change of chemistry composition and biochemical properties of nutrient media and environment. Change of type of optimal cultures is the result of metabolic shift of chemical composition and biochemical properties of nutrient media. The case b) presents the process of growth impairment in "one-line" culture at metabolic shift.

3.5.2. *Experiments on transmutation of isotopes in growing microbiological associations*

The general aim of investigation was connected with searching of optimal methods of biotechnology for effective isotope transmutation.

Expected $Mn^{55} + d^2 = Fe^{57}$ reaction with heavy water in growing microbiological associations was conducted in the system

"D^2O+Mn^{55} + MCT + additional isotope components".

During experiments microbiological MCT granules were used.

The control experiments were conducted in another system

"$H_2O + Mn^{55}$ +MCT + the same additional isotope components".

Many different experiments under varying conditions were carried out.

A typical series of experiments concerning nuclear transmutation of elements consisted in growing of certain microbiological culture in three disks simultaneously.

The first disk contained full-compounded heavy-hydrous (D_2O) nutrient medium with $MnSO_4$, the second one—also heavy-hydrous (D_2O) nutrient medium without $MnSO_4$, and the third one—light-hydrous (H_2O) full-compounded (with 0.05% of $MnSO_4$) nutrient medium.

Such series of experiments were held for MCT during 20 days at temperature 25°C.

After the completing of each series, the obtained biological substance was collected, cleaned in distilled H_2O water and dried. The dried substance in the form of a unstructured granules (like peat) was separated using non-iron containing instrument, ground to a powder and placed at the same quantity in Mossbauer spectrometer. The mass of investigated dried biological substance in all cases was about 0.3 g.

The results of combined (both Mossbauer and TIMS mass-spectroscopy) examinations of isotopes transmutation process in growing microbiological associations in the same iron-region of atomic masses ($54 < A < 58$) are presented below.

The results of Mossbauer investigation of both control and optimal dried biological substances are presented in Figs. 3.17 and 3.18. In these experiments different amplitudes of Mossbauer resonance ($\Delta J/J \approx 1.1\%$) and ($\Delta J/J \approx 3.1\%$) at the same masses of investigated dried biological substances were observed and measured. It is well known that the amplitude of Mossbauer resonance is proportional to the concentration of resonant nuclei (if this concentration is low).

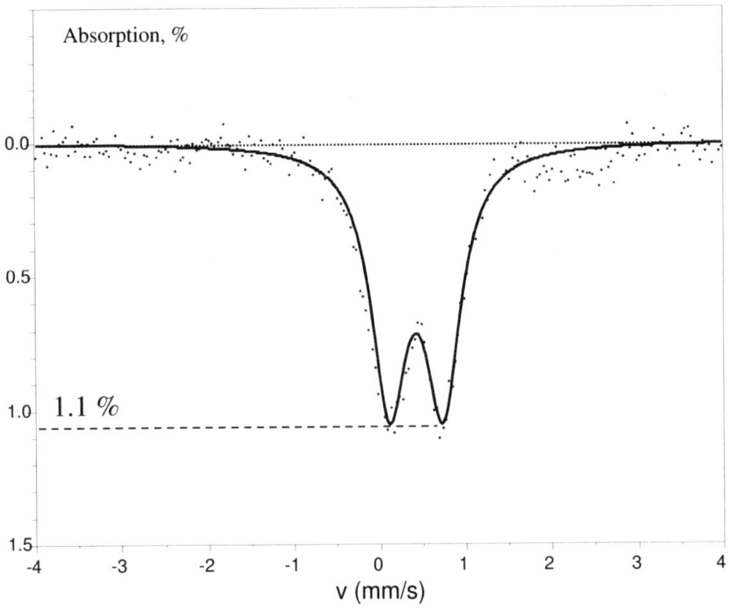

Fig. 3.17. Mossbauer spectrun of microbiological *MCT* grown in the volume with presence of H_2O and Mn^{55} isotope (control): $\Delta J_{max}/J \approx 1.1$ % is the magnitude of Mossbauer resonance.

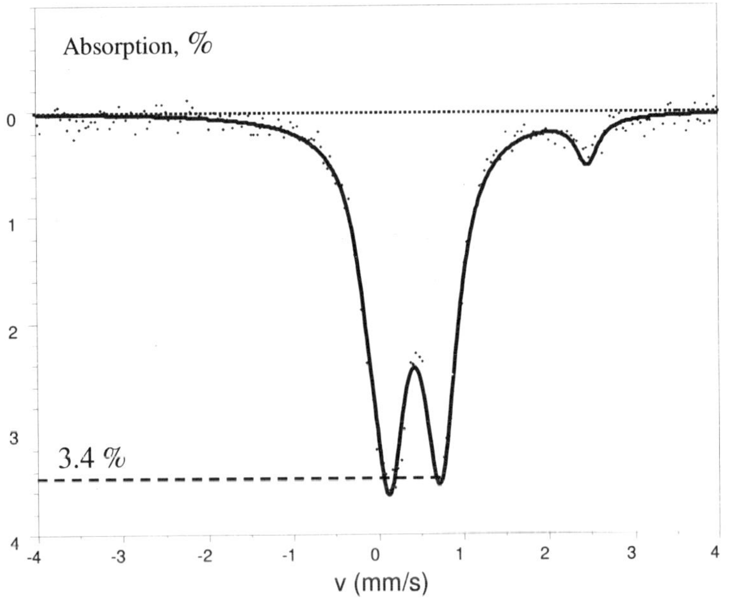

Fig. 3.18. Mossbauer spectrum of microbiological *MCT* grown in the volume with presence of D_2O and Mn^{55} isotope (experiments on transmutation): $\Delta J_{max}/J \approx 3.4$ % is the magnitude of Mossbauer resonance.

Registration of Mossbauer Fe^{57} isotope in the biological substance that was grown in nonoptimal medium (see Fig. 3.17 for the control experiment) is the result of presence of natural Fe as admixture in the components of initial MCT compound.

MCT compound is the special granules that include:

1. concentrated biomass of metabolically active microorganisms (microbe syntrophin association);
2. organic sources of carbon and energy, phosphorus, nitrogen, etc.;
3. gluing substances that keep all components in the form of granules stable in water solutions, for a long period of time, subjected to many, possibly any, external conditions.

The organic source of microelements contained some small quantity of natural Fe as admixture. In this case the registered Fe^{57} isotope is the small part (≈2.2%) of natural Fe.

In experiments on transmutation (see Fig. 3.18) the same MCT compound with the same organic sources of microelements was used. But the result (presence of Fe^{57} isotope) was different: concentration of Fe^{57} isotope in control biological substance less by three times in relation to transmutated biological substance.

So, the difference of quantity of Fe^{57} isotope in transmutated and control dried biological substances follows from the difference of Mossbauer absorption

$$(\Delta J_{max}/J)_{\text{transmut}} - (\Delta J_{max}/J)_{\text{control}} \approx 2\ \%,$$
$$(\Delta J_{max}/J)_{\text{transmut}} \,/\, (\Delta J_{max}/J)_{\text{control}} \approx 3$$

and is the direct result of fusion reaction $Mn^{55} + d^2 = Fe^{57}$ during growth of the same microbiology MCT compound in optimal medium.

The total, relative, number of Fe^{57} nuclei that was created is about 10^{17} nuclei per 1 g of grown and dried biological substances, which is between 10 and 20 times more than the comparable, relative, maximal number of Fe^{57} nuclei that is created in "one-line" grown and dried cultures. The total mass of Fe^{57} isotopes that is created is about 10^{-5} g per each g of dried biological substance. The efficiency has increased, in particular, because the association has been allowed to grow_during a 20 day period.

"One-line" cultures cannot be grown for such a long period of time in heavy water because of "self-intoxication" of the medium by the metabolic products (in our former experiments the "one-line" *Escherichia coli* culture was grown during a 72 hour period).

The relative efficiency rate λ of such forms of transmutation (the coefficient of transmutation) is the following:

$\lambda = \dfrac{N(Fe^{57})}{N(Mn^{55})\Delta t} \approx (0.5...1.0).10^{-6}$ (synthesized Fe^{57} nuclei per *s* and per single Mn^{55} nucleus).

For verification of these results additional examinations of isotopes ratio of the same dried biological substances (both control and transmutated) were conducted by TIMS (Thermal Ion Mass Spectroscopy, "Finnigan" MAT-262) spectroscopy. The results of TIMS measurements presented in Fig. 3.19 and in Table 3.8.

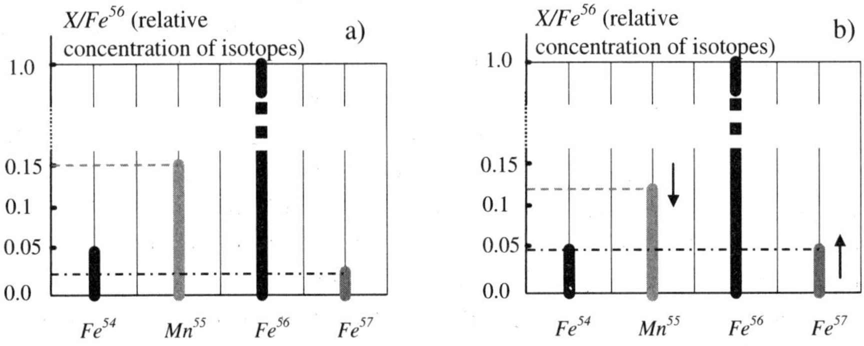

Fig. 3.19. Mass-spectrum of iron-region of microbiological associations (dried biological substances) that were grown in control nutrient medium with H_2O and Mn^{55} (case a)) and in experimental nutrient medium with D_2O and the same quantity of Mn^{55} isotope (case b)) Here $X = Fe^{54}; Mn^{55}; Fe^{57}$. The process of increasing (↑) of concentration of Fe^{57} isotope is accompanied by decreasing (↓) of concentration of Mn^{55} isotope

The amounts of Fe^{57} isotopes that are created are approximately the same in the cases of Mossbauer resonant gamma-spectroscopy and *TIMS* measurements (concentrations of Fe^{57} isotopes that are created increase by factors of 2...3).

The effectiveness of isotope transmutation during the process of growth of microbiological associations at optimal conditions increases by factors of 10- 20 times more than the effectiveness of the same transmutation in "One-line" (clean) microbiological cultures.

The structure and half-width of Mossbauer spectra of control and transmutated microbiological associations are identical. So, the process of transmutation does not appear to change the spatial structure of the growing biological culture. Created and natural Fe are identical in the biochemical sense!

Table 3.8. Parameters of mass-spectroscopy investigation of control and transmutated cultures.

Isotope, natural concentration	*Natural isotopic ratio (in relation to Fe^{56})*	*Concentration in dried biological substance in control experiment: $H_2O + MnSO_4$ + nutrient medium*	*Isotopic ratio in control biological substance*	*Concentration in dried biological substance in experiment on transmutation: $D_2O + MnSO_4$ + nutrient medium*	*Isotopic ratio in the experiment on transmutation*
Mn^{55}, 100%	–	0.15 ± 0.012	$Mn^{55}/Fe^{57} = 6.6$	0.13 ± 0.012	$Mn^{55}/Fe^{57} = 7.7$
Fe^{56}, 91.7%	1	1	1	1	1
Fe^{57}, 2.2 %	$Fe^{56}/Fe^{57} = 41.7$	0.024 ± 0.002	$Fe^{56}/Fe^{57} = 42.5$	0.051 ± 0.003	$Fe^{56}/Fe^{57} = 19.5$

Decreases in the amounts of the additional Mn^{55} isotope in the transmutation flask are synchronized with the creation of Fe^{57} isotopes in the same flask. This appears to provide proof (a "form of acknowledgement" or a "footprint") of nuclear synthesis in processes associated with a "growing," biological system!

3.6. Experiments on controlled nuclear transmutation and deactivation of long-lived radioactive isotopes in biological cells

3.6.1. *Accelerated deactivation of reactor water in biological cells*

Now in the world there are more than 100 thousand tons of spent reactor fuel (high-level radioactive waste). Besides, in each reactor there are about thousand tons of highly active water (about 1 million tons of highly active water in the world). Besides, in the world there are about 10 millions tons of low-level radioactive waste and different radioactive materials.

There are different possible methods of utilization of these waste.

Traditional way of possible utilization and deactivation of this waste is connected with their transmutation to different stable isotopes by action of fast neutrons produced in the system of numerous special high-current high-energy proton accelerators. This way is very expensive and nonoptimal. The total cost of scientific and technology parts of such utilization program for all "nuclear" countries (e.g., USA, Japan, UK, Russia, France, S.Korea) exceed $50 billions during 2010-2050 years!

Another essential drawback of this program is the following: at such neutron action on highly radioactive waste a great amount of additional low active waste is formed in environment.

We have studied the "natural" way of radioactive isotopes utilization that takes place during growth of microbiology systems (Vysotskii, Tashyrev, Kornilova, 2008).

In our experiments nuclear transmutation of several kinds of reactor radionuclides by MCT stable compound has been investigated.

The base of MCT are microbe syntrophin associations that contain many thousands kinds of different microorganisms that are in the state of complete symbiosis. Such symbiosis leads to the possibility of maximal adaptation of the microorganisms' association to any external conditions changes (including effect of highly active ionizing irradiation).

Microorganisms' interaction with metals is manifested through its final result as mobilization (transformation of metals into a dissolved state) or to the opposite as immobilization—accumulation (transformation of metals into a insoluble state or their tying by cells).

Apparently the radio nuclides just as non-radioactive metals will form sulphides of metals during biogenic sulphate reduction. In other words radioactive metals (radio nuclides) should be regarded first of all as elements reacting with microorganisms in the same manner and through the same mechanisms as non-radioactive metals.

Typical reaction of the association for such aggressive effects demands the existence of some time for internal adaptation. This time is necessary for mutagene change of ≈ 10 generations that corresponds to several days. During this time occurs a purposeful synergy process of stimulation of the mutant formation of such microorganisms, which are maximally adapted to the changed aggressive conditions. This MCT is able to develop actively, for example, in the water with very highly specific radioactivity, while ordinary, not radioactively stable, monocultures die in such environment very rapidly.

The research has been carried out on the basis of distilled water from first contour of water-water atomic reactor of Kiev Institute of Nuclear Research (Vysotskii, Shevel, 2003). The water with total activity about 10^{-4} Curie/L contained highly active isotopes (e.g., Na^{24}, K^{40}, Co^{60}, Sr^{91}, I^{131}, Xe^{135}, Ba^{140}, La^{140}, Ce^{141}, Np^{239}) was extracted from the active zone (see Fig.3.20).

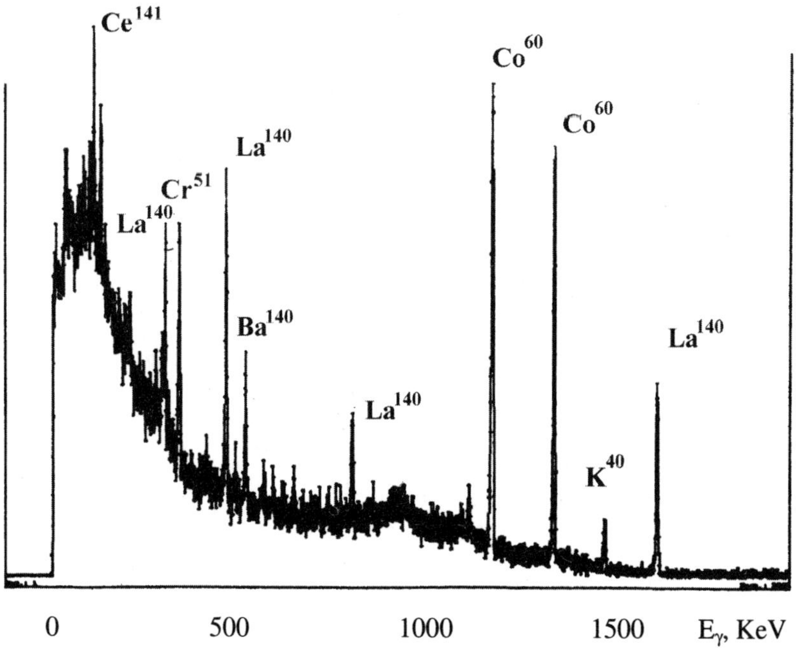

Fig. 3.20. Spectrum of gamma-radiation of radioactive distilled water from first contour of water-water atomic reactor (10th day after extraction from the active zone).

In our experiments MCT with mass about 1 g was placed in the planar glass flasks with 10 ml of water from the atomic reactor.

In control experiments the same radioactive water but without MCT was used.

The cultures were grown at the temperature 25°C. Activity of all flasks has been measured every 5 days by amplitude Ge detector of great size.

The results of investigation of activity Q(t) of the same reactor Ba^{140}, La^{140} and Co^{60} isotopes in the experiment on transmutation (activity is $Q_{cultures}$) and in the control one ($Q_{control}$) are presented on the Fig. 3.21. For the first time we have observed accelerated utilization of radioactive La^{140} and Ba^{140} isotopes in the flasks that contained "microbial transmutator" during all time of experiment (during 30 days)!

Studied La^{140} isotope has short life-time τ_{La} = 40.3 hours and is nonstable daughter isotope of Ba^{140} radioactive isotope that has life-time about τ_{Ba} = 12.7 days:

$$Ba^{140} \rightarrow La^{140} + \beta^-.$$

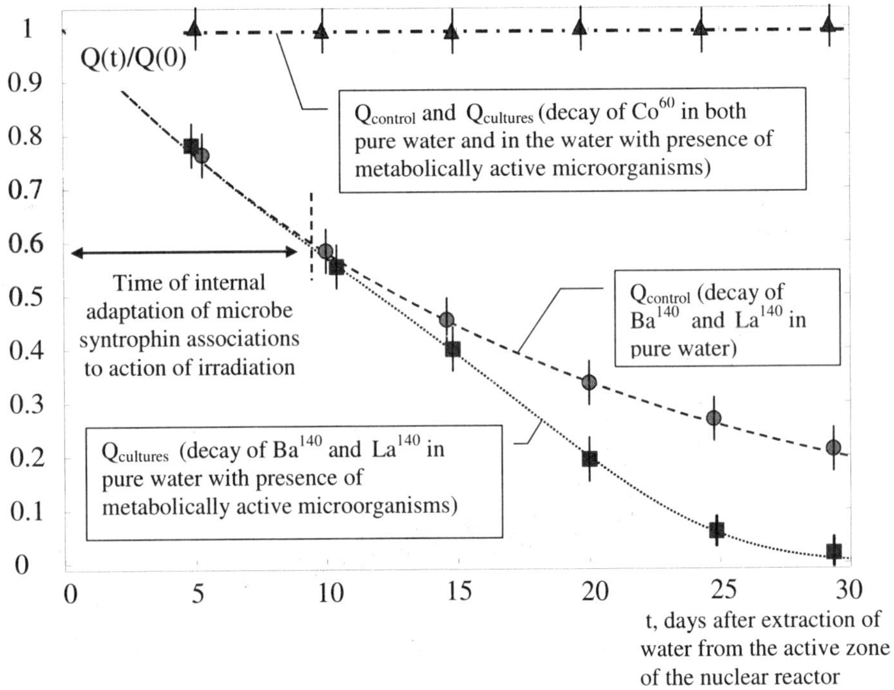

Fig. 3.21. Change of activity Q(t) of the same reactor Ba^{140}, La^{140} and Co^{60} isotopes in the experiment on transmutation (activity $Q_{cultures}$ in pure reactor water with presence of metabolically active microorganisms) and in the control one (activity $Q_{control}$ in the same pure reactor water without microorganisms)

Initial activities of the Ba^{140} and La^{140} isotopes (on the 10th day after extraction of water from the active zone of the nuclear reactor) were

$Q_{Ba\text{-}140} = 1.46.10^{-7}$ Curie/L and $Q_{La\text{-}140} = 2.31.10^{-7}$ Curie/L.

The possible way of radioactive Ba^{140} isotope transmutation to the stable state is

$$Ba^{140} + C^{12} = Sm^{152} + \Delta E$$

These reactions are energy favourable ($\Delta E = 8.5$ MeV is positive).

The Sm^{2+} and Ca^{2+} iones are chemically alike and have the approximately same ionic radiuses of divalent state ($R_{Sm} \approx 1.2$ A, $R_{Ca} \approx 1.06$ A). Substituted element Ca is among several vitally necessary elements. Ions of created Sm^{2+} elements can substitute Ca^{2+} ions while microbiological cultures are growing (see Chapter 5).

Probability of such substitution during process of growing of biological culture is high because the initial concentration of Ca element in MCT is low.

3.6.2. *Accelerated deactivation of long-lived active Cs^{137} isotope in microbe syntrophin associations*

It is well known, that the Cs^{137} isotope is the most dangerous long-lived component of spent reactor fuel. Investigation the possible ways of this isotope utilization and deactivation is very important.

The process of decontamination (deactivation) of radioactive waste through the action of growth in microbiological systems is connected with transmutation of long-lived active nuclei to different non-radioactive isotopes during growth and metabolic processes involving MCT granules.

The research has been carried out, based on using the identical distilled water but with a process that involves Cs^{137} with an activity of $Q=2.10^4$ bq (Vysotskii, Odintsov, 2004).

In the experiments, 8 identical closed glass flasks with very thin walls and with 10 ml of the same active water in each were used (see Fig. 3.22).

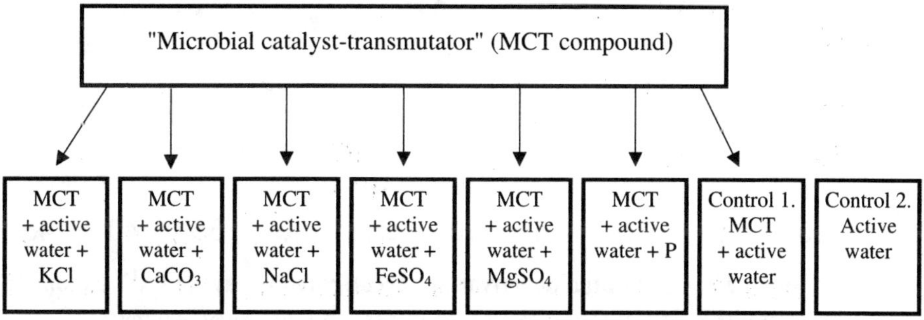

Fig. 3.22. Study of utilization of active isotopes at different conditions

The MCT compound was placed in 7 glass flasks. In six different flasks, different pure K, Ca, Na, Fe, Mg and P salts as single admixture were added to the active water. These chemical elements are vitally necessary for any cultures. Each of these specific replacements completely blocks all possible transmutation channels, in which any of biochemical analogs of the specific chemical element can be used.

We hypothesize that the reason these reactions can be blocked when these replacements are used is related to the formation of an optimal balance of microelements. The results obtained appear to confirm this hypothesis and, by implication, the importance of these kinds of replacements, as potentially blocking potential transmutations.

Two additional flasks were used for control experiments: one flask contained the active water and MCT (but without additional salts) and in another one was only active water (without salts and MCT).

The cultures were grown at the temperature 20℃. Activity of all closed flasks has been measured every 7 days by precise, large amplitudes, using a Ge detector. During the process of measuring of spectrum the special screened box with very low level of natural ionizing radiation background was used.

The results of investigation of change of relative activity Q(t)/Q(0) of isotopes are presented on Fig. 3.23 and Table 3.9.

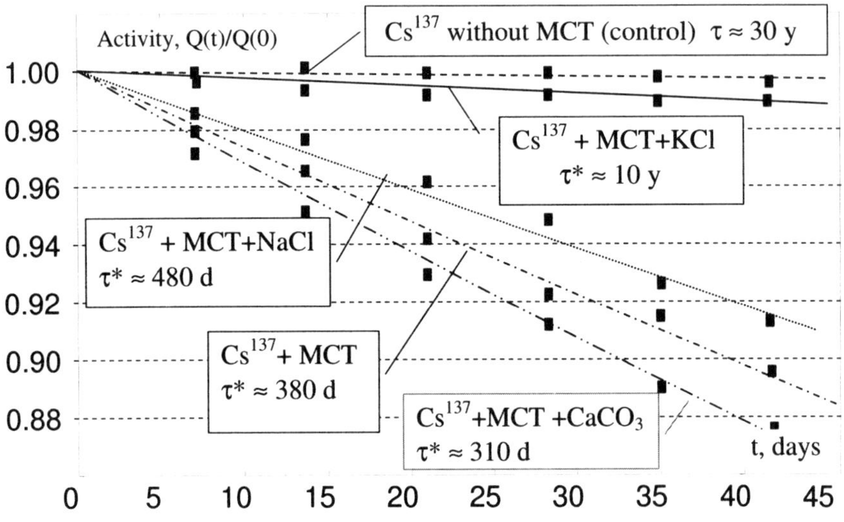

Fig. 3.23. Accelerated deactivation (accelerated decay) of Cs^{137} isotope in "biological cells" with presents of different chemical elements.

Table 3.9. Deactivation of different active isotopes in optimal experiment (MCT + active water with presence of Cs^{137}+ $CaCO_3$ solt)

Isotope; energy, keV	*Start*	*Finish of experiments (in 100 days)*			
	N_s, registered events per 10^3 s	*N_f, registered events per 10^3 s*	*Error (absolute/ relative)*	*Natural decay per 100 d*	*Change $(N_f T_s)/N_i$*
Cs^{137} 661.7	266900	216800	±478(±0.2%)	-0.6 %	-24 %

We have observed increased rates of decay of Cs^{137} isotope in all experiments with MCT and with the presence of different additional salts during more than 100 days.

In the control experiment (flask with active water but without MCT), the "usual" law of nuclear decay applies, and the life-time was about 30 years.

The most rapidly increasing decay rate, which occurred with a lifetime $\tau^* \approx 310$ days (involving an increase in rate, and decrease in lifetime by a factor of 35 times) was observed in the presence of Ca salt! In the presence of an abnormal (redundant) quantity of potassium in the nutritious media, the process of Cesium transmutation becomes very weak and the life-time of the decay was about 10 years.

A possible reaction of Cs^{137} isotope utilization is

$$Cs^{137} + p^{1} = Ba^{138} + \Delta E.$$

The result of this reaction is the creation of a stable Ba^{138} isotope. This reaction is energetically favourable ($\Delta E = 5.58$ MeV is positive).

The Ba^{2+} and K^{+} ions are chemically alike and have approximately the same ionic radii of the associated ionic state ($R_{Ba^{2+}} \approx 1.4A,\ R_{K^{+}} \approx 1.33A$).

We speculate that substitution of the element K can result in one, of among several, vitally necessary elements. Ba^{2+} ions can be created, in principle, (as in the last reaction) by substituting elements involving K^{+} ions in metabolic process while microbiological cultures are growing.

This substitution is potentially more effective than the "direct" replacement of Potassium for Cesium because the ionic radius of Cesium is $R_{Cs^{+}} \approx 1.65 - 1.69A$, which is larger than the ionic radius of $R_{K^{+}} \approx 1.33A$ of Potassium.

It was shown earlier in experiments with microculture *Blastocladiella emersonii* (Van Brunt, 1982) that the substitution of K^{+} ions for Rb^{+} and Ba^{2+} ions have taken place. These ions can replace each other in transporting ions through a membrane to a cell.

What is the reason for increasing the efficiency of transmutation by increasing the concentrations of calcium?

These phenomena are probably connected with general problems of metabolic processes involving microbiological cultures: optimal growth of microcultures takes place when a balanced relation of micro elements occurs.

The phenomenon of low energy transmutation of chemical elements and isotopes in biological systems and creating conditions for sustaining it is based upon the heuristic proposition that if some of the required elements or microelements are not present in the living environment (or nutrient media), then, given that certain pre-requisites are met, it will be synthesized as a result of the transmutation. In fact such an approach unambiguously suggests that the ratio of all the necessary elements in each type of living organisms is fixed.

These results reveal a non-trivial nature of interactions of different microelements.

By changing the makeup of the nutrient medium, it is possible to control the speed of a culture's growth. Lacking at least one of the microelements in the nutrient medium hinders the development of the entire biological object.

3.7. Investigation of memory phenomena in water and study of isotopes transmutation in growing biological systems containing activated water

3.7.1. *The problem of water memory*

The objectives of conducted investigations were to study the possibility and conditions of optimization of low-energy transmutation of different isotopes in growing microbiological cultures. The history of this optimization problem is connected with the several observed anomalies in our previous examinations of the effect of isotope transmutation in different living systems.

During numerious experiments on isotope transmutation in both pure microbiologilal cultures and microbe syntrophin associations we have observed abrupt and essential change of the effectiveness of the same nuclear reaction with the presence of the same cultures in flasks that were situated in different places of the same laboratory or in different times in the same place! It was mystery!

We suppose that these anomalies are connected with influence on water (on water-containing nutrient mediums) of random fluctuation of weak low-frequency electromagnetic background in different places of the same laboratory or in different times in the same place in the laboratory. The origin of these fields could be bound with influence of different outside laboratory equipment, which dies not concern these experiments.

In the process of vital activity of any biological object, the most important chemical compound is water.

It is well known that water:

- is a base of the intracellular liquid,

- forms the transport paths for the transfer of vitally necessary chemical elements,
- forms the necessary state of these elements in the form of atomic and molecular ions,
- play the main role at cell division and DNA replication;
- is the most universal solvent.

The spatial structure of water agrees ideally with the secondary structure of a DNA macromolecule, by guaranteing the maximally stable existence of the double helix formed of complementory pairs of nucleotides.

This list is far from being complete and can be continued.

At the same time, water is one of the most mysterious chemical compounds. Anomalous properties of water became long ago a classical example of the manifestations of characteristics of nontrivial systems.

For a long time, the continual (quasicrystalline) model of water was dominant. In the framework of this model, the spatial structure of the potential energy for each of the molecules H_2O is the almost periodic three-dimensional system of potential wells and barriers. This relief is a result of the self-consistent motion of all water molecules which combines two independent processes: the oscillatory motion in each of the potential wells and a random (fluctuation-related) hop to the neighboring well. The mean frequency of oscillations in potential wells is approximately the same as the Debye frequency in solids (about $\omega_D \approx 10^{13}$ sec^{-1}). The mean duration of a hop to the neighboring potential well $\tau_0 \approx 10^{-13}$ sec. The mean duration of the stay in a well,

$$<\tau>=\tau_0 e^{\Delta W/k_B T} \approx 10^{-9}-10^{-10} \text{ sec}$$

is determined by the temperature T of water and the activation energy, $\Delta W \approx 0.2$ eV, of the process of diffusion (by the height of a barrier between the neighboring wells). Staying in the framework of this model, it is easy to conclude that the water memory must be preserved not longer than the value of $<\tau>$ which is less by many orders than the values given by numerous experiments.

Up to now, there are no adequate answers to the questions about:

- spatial structure of water on molecular and supermolecular levels;
- existence of numerous paradoxes of water memory and the nature of water memory;
- duration of water memory and possibility of its control.

Some aspects of the problem of water memory (including calculation the memory duration) was studied in our works (Vysotskii, Kornilova, 2004; Vysotskii, Smirnov, Kornilova, 2005; Vysotskii, Kornilova, Smirnov, 2009).

Detailed studies showed that the so-called "clathrate" model is most close to the reality. In the final form, this model was developed by Pauling (Pauling, 1959). The Pauling clathrate model is based on the idea of that the union of atoms of oxygen and hydrogen is able to create spatial flexible tetrahedral frames. The Pauling clathrate model explains very well all the properties of water (including its anomalous compressibility). The structure of DNA perfectly corresponds to the spatial structure of such frame water. In this case, every macromolecule of DNA puts water in order at the distance to 300-500 Å from its surface. The possibility to join the Pauling clathrate model with the cluster model was considered in many works. In this case, the separate elements of clathrate frames can be joined from time to time by hydrogen bonds and form groups possessing an ordered structure (i.e., clusters). Since there exists a very strong interconnection between the neighboring hydrogen bonds, the appearance and elimination of hydrogen bonds occur in a correlated manner and synchronously in time. Such a character of the bond allows us to suppose that the "flickering clusters" arise and disappear in water. The lifetime of clusters is of order of 10^{-10} sec, i.e., of order of 1000 molecular oscillations.

The considered specific features of the bulk water structure indicate that water molecules are always distributed between two systems weakly coupled with each other: the quasiamorphous nonstructurized water and the quasicrystalline structurized system of clathrate hydrates. During the external action on water (i.e. on the activation of water), there occurs a significant change of its structure and parameters. In view of the scale and mechanism of activation, можно выделить two different hierarchical levels of organization of the water structure (a macrolevel and a microlevel).

The first hierarchical level of the water structure (the macrolevel of the structurc) corresponds to the global spatial structure of water and defines the form and the position of its spatial frame. This level is characterized by the presence of the system of clathrate hydrates which form stable dodecahedral polyhedrons of ions of oxygen and hydrogen. Inside the volume of each of these polyhedrons, there exist the empty microcavities with rigid hydrophobic walls. The dodecahedral polyhedrons with the help of stable hydrogen bonds are joined in in binary, triple, and more complicated associates which be combined in very large associates (macroclusters). The space between macroclusters is filled by the quasiamorphous water.

Thus, the macrolevel of the structural organization of water corresponds to the equilibrium distribution between the phase of amorphous water and the phase of water represented as a system of macroclusters with a complicated hierarchy. Under the action of the external factors, this distribution can be changed. For example, the volume of macroclusters increases with decrease in temperature, whereas the volume of quasiamorphous water decreases. As the temperature grows, the volumes of macroclusters decrease, and, in addition, each macrocluster can be divided in several less parts. It is natural that the volume of quasiamorphous water increases in this case. The same changes can occur under other types of action (for example, under the action of ultrasound on the aqueous medium). Due to the strong dependence on external actions, the macrolevel of the water structure has no sufficient efficiency in order to organize a water memory system stable to the external destructive actions. At the same time, it is obvious that, in the absence of very strong destructive actions, the process of recording of the information in the form of the system of regularly joined clathrate cells is quite possible. Simply saying, separate polyhedrons of the clathrate frame can be combined by several alternative means. In this case, of importance is the circumstance that the realization of a certain orientation of a specific pair of polyhedrons leads automatically to that the subsequent polyhedrons will be joined to this "bare" associate in exactly the same way, which induces the appearance of ordered macroclusters. Such a system has obvious long-range order, which can explain the possibility of a global structurization of great volumes of water.

The second hierarchical level of the water structure (microlevel) corresponds to the processes of motion and distribution of separate molecules H_2O between microcavities of the three-dimensional clathrate frame of water and the quasiamorphous nonstructurized water. This microlevel defines the nonstationary evolution of molecules H_2O. The process of evolution is determined by two possible directions: molecules can leave the volume of quasiamorphous water, enter into the volume of these microcavities, and be there in the hydrophobic form for a long time or, on the contrary, can pass from microcavities into the volume of the quasiamorphous water.

For the alternative direction of the relaxation (the transition of a water molecule from the volume of a microcavity into the volume of amorphous water), the duration of relaxation T_{2W} is determined also by a relation of the type

The results of calculations (Vysotskii, Kornilova, 2004;Vysotskii, Smirnov, Kornilova, 2005) of the durations of two type of relaxation (T_{1W} for transitions of a water molecule H_2O from the volume of amorphous water into the microcavity and T_{2W} for transitions of a water molecule from the volume of a microcavity into the volume of amorphous water) are presented in Table 3.10.

Table 3.10. Dependence of the duration of relaxation of water (the duration of "the water memory") on its temperature

T, °C	1	10	20	30	36.6	40	50	60	70	90
T_{1W}	300 days	49 days	10 days	58 h	24 h	15 h	4.4 h	1.3 h	27 min	3 min
T_{2W}	30 min	14 min	4 min	1.5 min	45 sec	30 sec	12 sec	4 sec	1.5 sec	0.3 sec

We performed a cycle of complex studies of activated water and water memory, by using the activator of water, which was earlier studied by us (Vysotskii, Kornilova, 2004;Vysotskii, Smirnov, Kornilova, 2005) and allowed us to obtain a number of results in the whole spectrum of physical, biological and nuclear applications. We have conducted detailed studies of mechanical, electrodynamic, optical, and other characteristics of activated water itself and various physical, biological and medical phenomena which run with its participation (Vysotskii, Olishevsky, 2006; Vysotskii, Tashyrev, 2006; Vysotskii, Kornilova, Smirnov, 2009). The part of these results are presented below.

3.7.2. *Investigation of physical-molecular properties of activated water*

Water was activated by the action of nonionizing Molecular Resonance Effect Technology (MRET) (Vysotskii, Smirnov, Kornilova, 2005; Vysotskii, Kornilova, Smirnov, 2009). Investigated activator of water is the stationary source of low-frequency resonant electromagnetic field with composite space structure and very weak amplitude. The device includes the working body consisting of a polymeric matrix formed by oriented threads of a linear polymer with the alloying admixtures of various chemical elements and compounds. The polymer has the properties of both ferroelectric and piezoelectric. This polymer is surrounded by the system of mutually perpendicular pairs of oriented constant magnets positioned in turn, the field intensity of each magnet being 4000 *Oe*. Periodic optical pulses (frequency is about 7-8 Hz) acting on the oriented polymeric system. The motion of the electric charges and the magnetic moments of atoms in the strong circular field of the constant magnets generates a variable low-intense electromagnetic field (about 1 *Oe*) with complicated spatial structure. This field oscillates with the frequency of the repetition of optical exciting pulses and includes both the electric and magnetic components.

The generated electromagnetic field acts on water positioned near the open end of the polymeric activator and changes its structure, which is reflected in the long-term change of the properties of water.

It was discovered in our detailed physical experiments that under the action of this irradiation there are very essential modifications of the basic physical-molecular and biophysical properties of distilled water.

3.7.2.1. Anomalous electrodynamic characteristics of activated water

In this section, we consider the main electrodynamic characteristics of activated water.

In the process of studies, we determined the electrodynamic properties of activated water as functions of three basic parameters:

- the water activation duration,
- the water storage duration after the activation,
- the temperature of activated water storage.

The influence of some of these parameters was investigated comprehensively (for example, the influence of the storage duration and the storage temperature after the activation).

In Fig. 3.24a the results of the base studies of the electrodynamic characteristics of initial nonactivated distillated water completely analogous to water which was then subjected to the procedure of activation are presented. The studies were carried out at a temperature of 20°C.

It follows from these results that the initial distillated water has a low value of conductivity, $\sigma \approx 3.10^{-5}$ S/cm^2, and the dielectric permittivity close to the "standard" one, $\varepsilon'(\omega) \approx 90$, at the frequency of alternating electric current in the range $\omega \geq 10^3$-10^7 Hz. This value of the dielectric permittivity well agrees with the Debye mechanism of orientational polarizability of independent molecules of water. The maximum value of the dielectric loss tangent $(tg\delta)_{max} \approx 120$ and corresponds to a frequency of 1522 Hz.

With decrease in the frequency to $\omega < 10^2$ Hz, a sharp decrease of the conductivity and a very sharp monotonous increase of the dielectric permittivity were registered. In the studied region of superlow frequencies (at $\omega \leq 10^{-1}$ Hz), the effective dielectric permittivity reaches a very great value $\varepsilon'(\omega) \approx 10^8$.

In Fig. 3.24b and Fig.3.24c we present the specific features of the electrodynamic characteristics of water activated for 30 min and stored up to the time moment of the measurement at a temperature of 20°C and 40°C. It is seen that these samples of water preserves some general regularities (a sharp decrease of the dielectric permittivity and the conductivity on the activation of water and their gradual recovery in the process of storage) and, at the same time, reveals the significant differences such as a significantly faster relaxation of the parameters with increase of the storage duration. In particular, for 5 h of the storage, the maximum value of the dielectric permittivity and the conductivity grew by 200% (see Fig 3.24b).

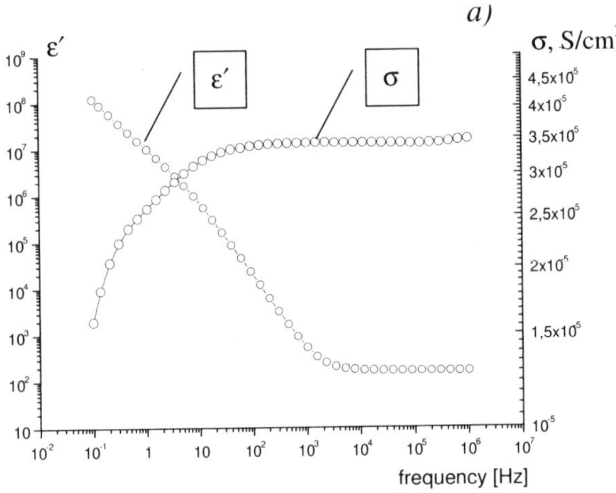

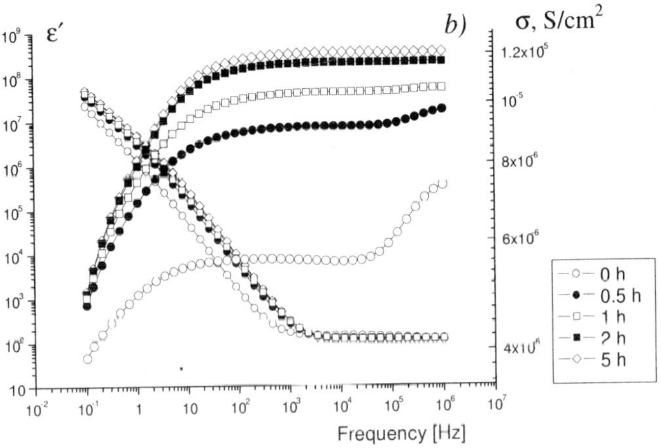

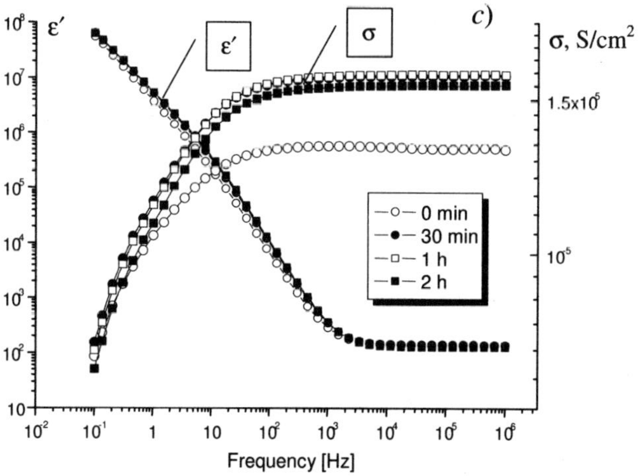

Fig. 3.24. Dielectric permittivity ε′ and conductivity σ of a water versus the frequency: *a*) initial nonactivated distilled water at 20°C; *b)* the same distilled water activated for 30 min and stored at a temperature of 20°C; *c)* the same water activated for 30 min and stored at a temperature of 40°C.

One of the most important problems is the determination of the duration of "the memory of water" on the basis of the performed experiments.

In Fig. 3.25, we present the dependence of the relaxation change of $\varepsilon'(\omega, t)$ on the storage duration of activated water after the activation.

These results correspond to the frequency $\omega = 10$ Hz and are obtained on the direct processing of the above-discussed spectra obtained for water stored for the different time intervals at different temperatures after the activation. Water was activated for 30 min.

It is seen that, at very great storage duration, the anomalous electrodynamics properties of activated water gradually disappear. The duration of preservation of these properties depends very significantly on the storage temperature.

By extrapolating the obtained dependences of the relaxation change of the parameters of water $\varepsilon'(\omega)$ and $\sigma(\omega)$ by the exponential law

$$\varepsilon'(\omega,t) \approx \varepsilon'(\omega,\infty) - \Delta\varepsilon'(\omega)\exp(-t/T_W),$$
$$\sigma(\omega,t) \approx \sigma(\omega,\infty) - \Delta\sigma(\omega)\exp(-t/T_W),$$

we can estimate the duration of relaxation T_W.

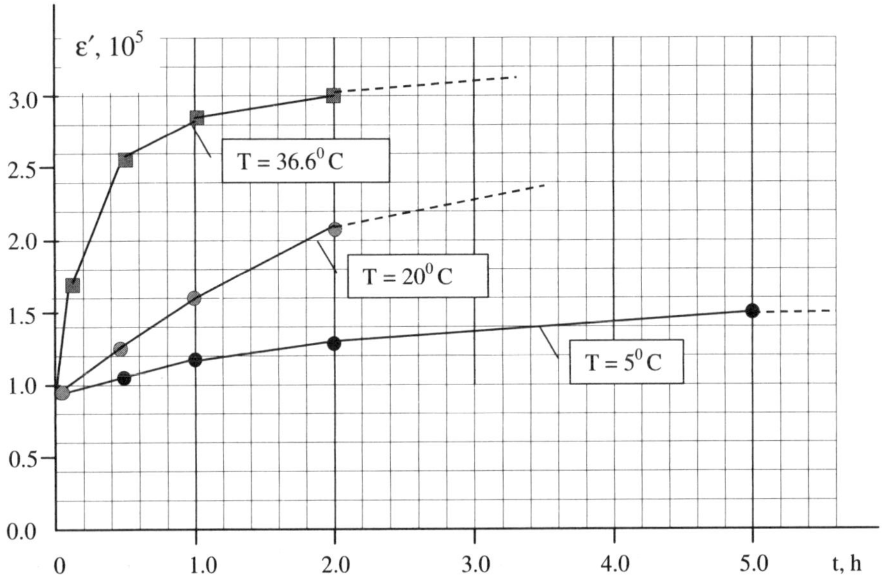

Fig. 3.25. Relaxation of the dielectric permittivity of activated water $\varepsilon'(\omega)$ at the frequency $f = 10$ Hz ($\omega = 2\pi f \approx 62$ sec^{-1}) versus the storage duration after the activation. Three curves correspond to three temperatures of the storage of water (T =5°C, 20°C, and 36.6°C).

It is seen from Fig. 3.25 that the duration of relaxation T_W of the mentioned electrodynamic characteristics of activated water becomes very great (at least 5-7 days) at a comparatively low temperature (at 5°C). With increase of the temperature of the storage, this duration rapidly decreases. The duration of relaxation is 10-15 h at a temperature of 20°C and is equal to at most 4-6 h at a temperature of 40°C close to the temperature of a human body. These results qualitatively agree with the calculations of water memory performed in our works (Vysotskii, Kornilova, 2004).

3.7.2.2. Anomalous viscosity characteristics of activated water

Viscosity is one of the most important characteristics of water. The particular meaning of the viscosity of water in any physical and biophysical systems is related to the transport functions of water and to the dependence of the motion and the evolution of macromolecules, ions, cells, viruses, and other microobjects on the properties of water, in which they are placed in alive organisms.

In our experiments, the viscosity of water was measured with the help of a rheometer. The studies were executed on the basis of a measuring cell. It consisted of two immovable 1 and 2. In the region between them, there is a rotor (a rotating coaxial cylinder 3 coaxial with cylinders 1 and 2). Water under study is placed in the free region between the surfaces of immovable cylinders 1 and 2 and the rotor. With the help of this system of coaxial cylinders, we performed the measurements at low shear stresses.

To study the influence of the activation of water on its viscosity, we executed several series of measurements of the dynamical viscosity η as a function of the applied stress τ with different fractions of water (the initial nonactivated distillated water and an analogous water activated for different time intervals) at two temperatures (20°C and 36.6°C).

In Fig. 3.26 we present the results of studies of the viscosity coefficient for initial distillated nonactivated water and for the samples of water which were obtained for the duration of activation of 15, 30, 45, and 60 min and were at a temperature of 36.6°C in the process of measurement. With regard for the fact that the anomalous properties of activated water relax rapidly with increase of the temperature, the measurements were carried out at once after the activation of water.

It was discovered that at very low velocity of motion of investigated activated water (tension less than 0.01 Pa) viscosity of water, activated during 60 min, decrease by 80-100 times at 20°C in relation to the same but nonactivated water. Minimal viscosity at such temperature will be for water activated during 30 min. In this case viscosity decrease by 300 times! It is seen from the obtained data that an increase of the water activation duration leads to a very significant decrease of the viscosity coefficient in the region of small mechanical stresses at this temperature (relative to the initial nonactivated water).

It is necessary to note that the absolute values of the viscosities of nonactivated and activated water samples in the region of small mechanical stresses increase significantly with the temperature.

It was discovered that duration of internal storage of these abnormal characteristics of activated water equals several days or weeks at low temperature (close to 0°C). The problem and concrete mechanism of long time memory of activated water were studied earlier in (Vysotskii, Kornilova, 2004; Vysotskii, Kornilova, Smirnov, 2009).

The change in the dispersive electrodynamic properties of water can render a very strong influence (by means of a modification of the electrostatic forces between separated charges and forces of the van der Waals type defining the interaction of the systems of neutral atoms and molecules) on the long-range interaction of such basis elements of alive systems as cells, viruses, biological macromolecules, enzymes, etc.

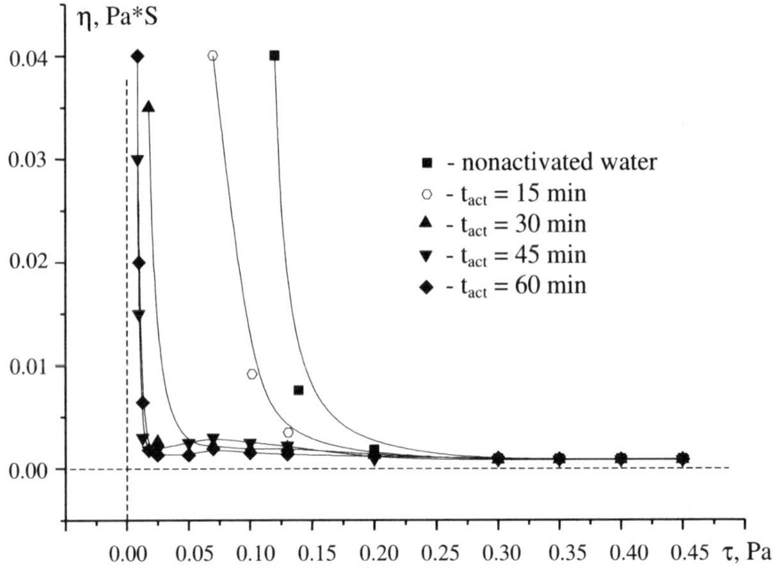

Fig. 3.26. Viscosity η of nonactivated water and four fractions of activated water in the region of small shear stresses τ at a temperature of 36.6°C.

Viscosity, surface tension, and other mechanical properties of water which play a very important role in the processes of vital activity turn out to be directly related to the electrodynamic characteristics of water.

It is seen from Fig. 3.26 that a change (increase) of the duration of activation in the limits of 1 h leads to a monotonous displacement of the curve of the viscosity coefficient as a function of the applied mechanical stress to the side of less viscosities. The position of the minimum of the viscosity coefficient is also shifted with increase of the duration of activation from $\tau \approx 0.1$ Pa for the duration of activation of 30 min to $\tau \approx$ 0.05-0.02 Pa for the duration of activation of 45-60 min. Very important is the circumstance that the viscosity of activated water in the region of very small shear stresses at the vitally important temperature 36.6°C, like the case of a lower temperature of 20°C, is by several orders less than that of ordinary (non-activated) water.

3.7.2.3. Influence of the activation of water on pH

It is well known that a value of the hydrogen index pH of the intracellular and intercellular water-salt medium is one of the essential factors of the influence on the course of biochemical processes and the development of alive organisms. With the purpose to study the influence of the activation of water on pH, we studied the dependence of pH on the duration of activation of water (water was activated for 15, 30, 45, and 60 min), on the temperature (at 4°C and 20°C), and the duration of storage of this water after the completion of the activation.

In experiments, we use water obtained on the distillation directly prior to the studies. Then the 100-ml samples of activated water were stored in standard graduated flasks of 250 ml in volume closed by rubber plugs. The first 3.5 h after the activation, all samples including the control ones were at room temperature (20°C). Then, according to the conditions of the experiment, a half of flasks was placed into a cooler and was stored there at 4°C, and the other half remained at 20°C. The control samples with analogous, but nonactivated water were present in both versions.

Each measurement of water activated for a definite time interval was repeated three times. Then we determined the mean value. The final error of measurements in each series was at most $\Delta(pH) \approx 0.1$.

The detailed studies of all the samples of water obtained for the different durations of activation were performed with a short time interval (in each 30 min) for the first 3.5 h after the completion of its activation at a temperature of water 20°C. The subsequent measurements were performed regularly for 16 days with an interval of 1-2 days for different fractions of water stored at temperatures of 4°C and 20°C. The samples of water taken from a cooler, where they were stored at a temperature of 4°C, were placed prior to the measurement in a thermostat, where their temperature was rapidly increased to 20°C. Then we measured the hydrogen index. The results of measurements are presented in Figs. 3.27 and 3.28.

In addition, the separate measurements were carried out by using the samples of nonactivated water and activated water stored after the activation at a temperature of 4°C for 120 days.

It was established that a quite insignificant alkalization of water occurred during the direct action of non-ionizing activator on water. In particular, when the hydrogen index for the initial distillated nonactivated water $(pH)_0 = 6.28$, a value of pH at once after the completion of the activation for 15, 30, 45, and 60 min was equal to $(pH)_{15} = 6.29$, $(pH)_{30} = 6.31$, $(pH)_{45} = 6.33$, $(pH)_{60} = 6.30$. The further studies showed that the process of alkalization of activated water became gradually more intense for the first 1.5 – 2 h of its storage after the activation. These data are presented in Fig. 3.27.

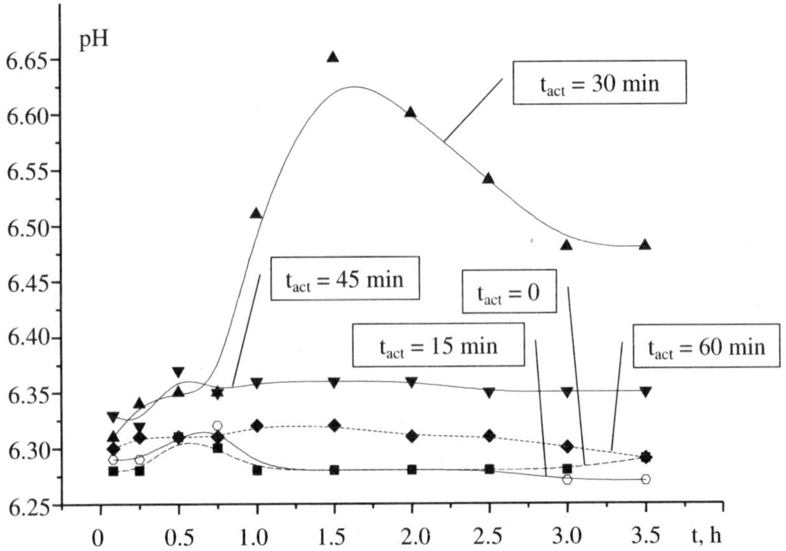

Fig. 3.27. Hydrogen index pH of different fractions of activated water versus the duration of its storage at a temperature of 20°C in the first hours after the activation.

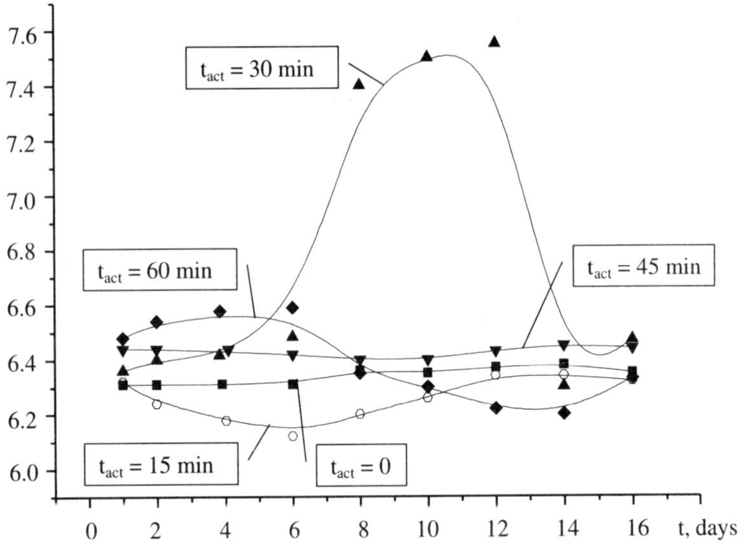

Fig. 3.28. Influence of the water activation duration on pH depending of the time of its storage at a temperature of 20°C.

The effect of the increase in pH is especially clearly manifested for water activated for 30 min. In this case, a value of pH increased from pH=6.31 to pH=6.65 during 1.5 h after the completion of the activation and then decreased gradually for 2 h down to pH=6.45. A small increase of pH was also observed for water activated for 45 min. For the rest fractions of activated water, the changes for the first hours of the storage were insignificant.

The study of the samples of activated water and control one at a longer time interval of the storage at two temperatures (4°C and 20°C) showed that the anomaly in the behavior of pH observed directly after the activation of water during 30 min was not unitary. In the process of storage, we registered the additional spontaneous increase of pH to a value exceeding the first maximum. The greatest changes in pH were discovered in the same fraction of water in 9-11 days of the storage at various temperatures. In particular, in 9 days of the storage, a value of pH grew to pH =7.4-7.5 [as compared with pH= 6.30-6.36 in control samples of nonactivated water which were stored at the same temperatures 4°C and 20°C. A half-width of this maximum (the duration of its existence) corresponded to $\Delta t \approx 4$ days at a water storage temperature of 4°C and $\Delta t \approx 6$ days при 20°C. Thereafter, there occurred a spontaneous decrease of pH which "returned" to its initial value, pH ≈ 6.3-6.4, only in 14 days. This maximum was observed on all samples of water stored and measured separately which correspond to the duration of activation of 30 min and the storage temperatures of 4°C and 20°C. There are the weighty reasons suppose that similar oscillations can exist also for a greater duration of the storage of water.

We note that the fluctuations of pH for control (nonactivated) water stored under the same conditions were practically invariable (with regard for the errors of measurements).

It is worth noting the essential fact that the displacement of pH to the alkaline side was fixed also at a long-term (for 120 days) storage of the samples of activated and control water. All these samples were obtained at the same time on the basis of identical distillated water.

In conclusion, we note that the presence of great ordered spontaneous changes in pH for the great duration of storage of water activated in a certain manner testifies directly to the existence of the "memory" of water. The presence of such great ordered changes in the properties of activated water is related, most likely, to certain phase transitions. Very curious is the fact that the time moment of the realization of these phase transitions, as well as the amplitude of changes in pH, does not depend (or weakly depends) on the temperature of the storage of activated water and is approximately equal to 10-11 days. At the same time, the half-width of the region of this spontaneous change depends rather sharply on the temperature.

3.7.3. *Investigation of inhibition biological activity of activated water*

The examination of influence of activated water on growth of strain *Escherichia coli* microbiological cultures on nutrient medium meat broth with 1.5% agar was carried out.

The study of the cultural-physiological parameters of culture *Escherichia coli* during its growth on the activated and nonactivated agarized medium was made with the purpose to determine the effect of the activation on the survivability of cells, the growth rate of colonies on the base of survived cells, and the shape and the size of grown cells. Cultivation of colonies was produced at 20°C in aerobic conditions on the surface of agar during 32 hours.

The cultural-morphological properties of culture *Escherichia coli* (the size, shape, and color of colonies and cells) are the important criteria of the effect of activated water on microorganisms. The change of the mentioned parameters evidences for influence of the activation on the cell division processes, the synthesis of the structural components of cells, pigments, the "programming" of the formation of a stereometry of colonies, etc.

Initial view of Petri dishes with different variants of activated medium at the beginning of experiments and the general view in 7 hours after beginning of experiments are presented in Fig. 3.29. The general view of the dishes with different fractions of activated nutrient medium in 29 hours after beginning of experiments is shown in Fig. 3.30. It was observed for the first time that at low initial concentration of cells of investigated culture *Escherichia coli* in nutrient medium water activated during 0.5 hour and 1 hour have suppressed culture growth in $N_C/N_{0.5}$ = 27 and $N_C/N_{1.0}$ = 303 times. At the increasing of water activation time, inhibition of conditionally pathohenic strain growth is more effective. In this case, activated water leads to inhibition of *Escherichia coli* i.e. decreasing of rate of growth.

The strong influence of activated water on the process of cloning of microorganisms, their division, the size of microorganisms' colony and the form of cell-like division has been discovered. It was observed during direct microscopy investigation that one of the reasons of abnormal growth is connected with suppressed division of cells in activated nutrient medium.

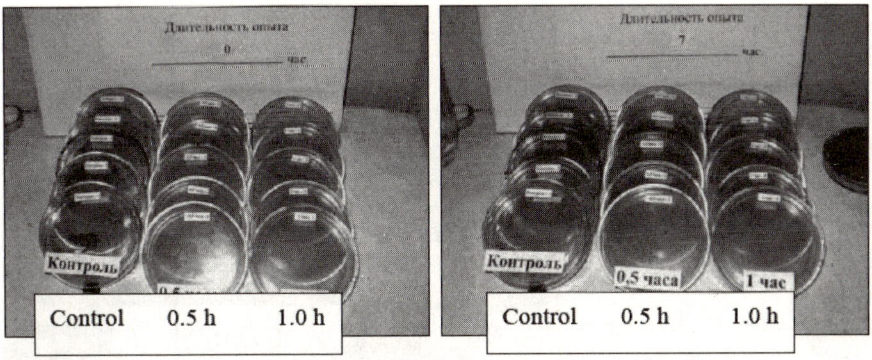

Fig. 3.29. General view of Petri dishes at the beginning of experiments (left) and in 7 hours after beginning of experiments (right). Identical very small amount of *Escherichia coli* cells was inoculated on a surface of three fractions of activated nutrient medium (t_{act}=0.5 h, t_{act}= 1.0 h and control). Colonies of microorganisms are absent in all dishes.

The anomaly of the cell division is manifested in this case through the creation in the colonies of *Escherichia coli* culture of the so-called "squibs" which are "blown" cells, whose division is violated but the growth is preserved. In the case of activated media we have observed many of cells, which after completion of growth are not parted after division and the linear line-ups consisting from 2-3 sequentially or more of paired cells form.

In Fig. 3.31 we present the photos of fragments of the system of cells that were grown on the surfaces of the nonactivated and activated agarized nutrient media. These fragments correspond to Fig. 3.30.

As seen from the fragment shown in Fig. 3.31, the main part of cells on the surface of the nonactivated medium corresponds to the shape "standard" for culture *Escherichia coli* (like elongate linear sticks), and the number of anomalous (nonseparated) cells is few. The estimations given above show that the concentration of anomalous cells is at most 4%, which is within the statistic fluctuations.

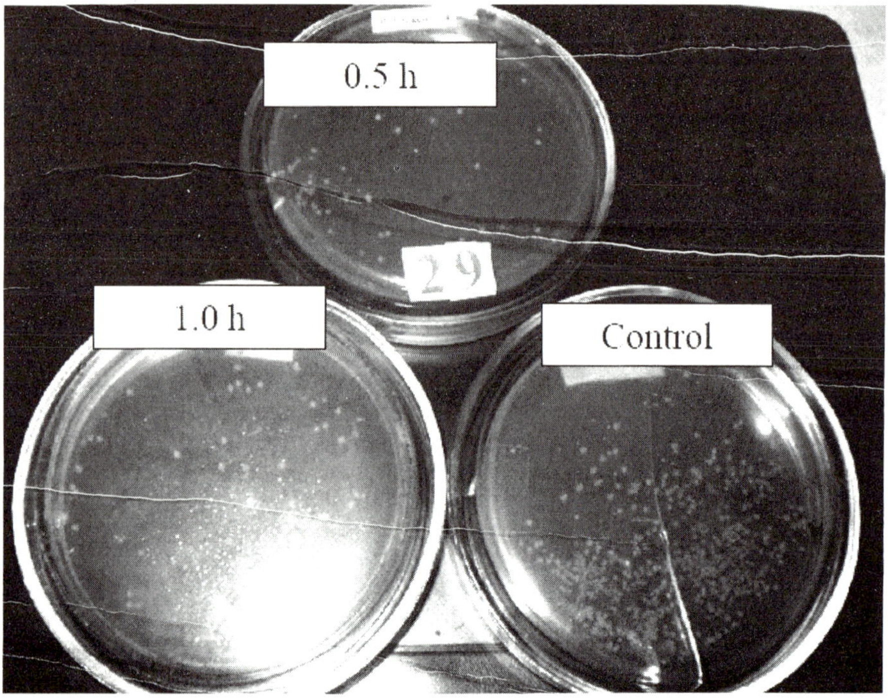

Fig. 3.30. View of the most representative Petri dishes with grown colonies of *Escherichia coli K-12* in 29 hours after beginning of experiments. a) Control. Number of cells is $N_C = 1.7.10^8$ cells/ml. Averaged diameter of grown colonies is d = 1.2 mm; b) $t_{act} = 0.5$ h. Number of cells is $N_{0.5} = 6.4.10^6$ cells/ml; c) $t_{act} = 1.0$ h. Number of cells is $N_{1.0} = 5.6.10^5$ cells/ml.

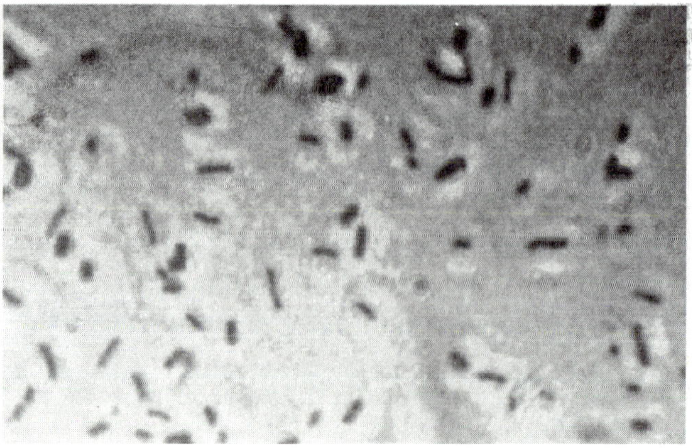

Fig. 3.31. Microscopic photos of cells of *Escherichia coli* grown on

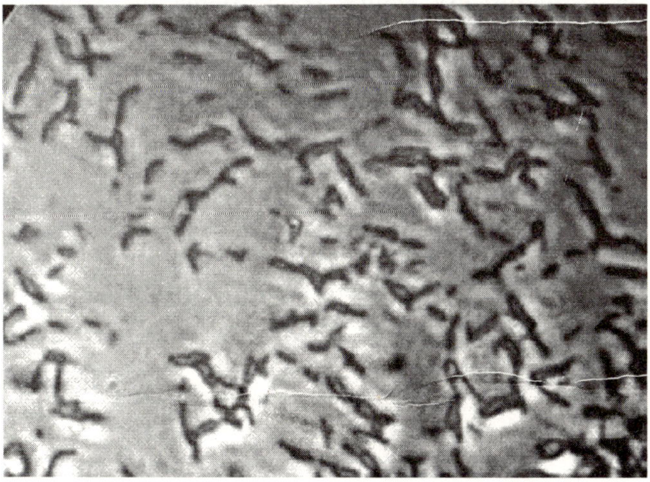

nonactivated medium (upper photo) and on the same but activated medium with t_{act} = 1 hour (lower photo).

The basically different pattern of the morphology of cells is observed for the culture grown on the activated medium. As follows from the magnified fragment of the surface of such nutrient medium shown in Fig. 3.31, the anomalous cells become the dominating component of the process of division of cells. Their relative concentration exceeds that of normal cells by a factor of 2.7. It is worth to note that, in this case, not only nonseparated linear chains containing 2-3 cells are formed. We observe also the great number of nonseparated branched chains of cells with a complicated branched structure that contains many (up to 10-20) nonseparated cells in some cases. With regard for the fact that the cell division process in such anomalous systems is to be suppressed severely, one of the natural reasons for the formation of these superstructures is, apparently, the mutual joining of individual linear nonseparated fragments. It is seen that, although the cell division process occurs most likely in accordance with the natural laws, the division rate of cells is sharply decreased on the last stage of their growth. It seems almost evident that just the change of the properties of activated water plays the key role on this last stage.

This is conditioned by the fact that just the energetic characteristics of the cell division process, which leads to the increase of both the surface and the total surface energy of dividing cells, are directly connected with the influence of the "external" (with respect to cells) water which is the base of the nutrient medium on their surface energy. The results of bactericidal (bacteriostatic) action of activated water on culture *Escherichia coli* growth are shown on Fig. 3.32.

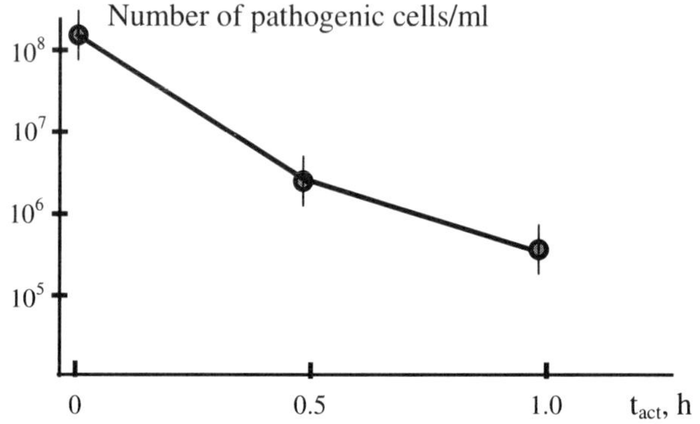

Fig. 3.32. Bactericidal (bacteriostatic) action of different fraction of activated water on culture *Escherichia coli* growing under aerobic conditions

The biophysical reasons of strong action of activated medium on the process of cloning and growth of pathogenic microorganisms are connected with:

- abnormal physical properties (viscosity, conductivity and permittivity) of activated water (activated medium);
- influence of such water (medium) during process of cells division on the balance of forces in the area of a cell walls;
- abnormal dynamics of the process of cells division in a volume of activated medium.

Activated water renders the bactericidal (or bacteriostatic) action on culture *Escherichia coli* growing under aerobic conditions. It is necessary to expect a similar influence under the action on other pathogenic cultures (for example, activated water should suppress the development of conditionally pathogenic bacteria *Citrobacter*, etc.).

3.7.4. *Transmutation of intermediate mass isotopes in growing biological cultures in activated water*

Our experiments on nuclear transmutation of intermediate mass isotopes in activated water were based on the expected

$$Na^{23}+P^{31}=Fe^{54}$$

reaction in growing culture in sugar-salt nutrient medium deficient in Fe but containing controlled quantity of Na^{23} and P^{31} isotopes. The research was carried out on the basis of the same bacterial culture *Escherichia coli*. The composition of the nutrient medium is presented in Table 3.11.

Table 3.11. The composition of the nutrient medium

Components	*Concentration in the medium (%)*	*Admixture of Fe (no more) relative (%)*
Sucrose	2	10^{-4}
$MgSO_4$	0,05	2.10^{-4}
$CaCO_3$	0,2	$1,5.10^{-4}$
KCl	0,05	3.10^{-4}
$NaNO_3$	0,5	2.10^{-4}
K_2HPO_4 (in transmutation experiments)	0,2	5.10^{-4}
Pure distilled water (activated or nonactivated H_2O)	100 (100 ml)	10^{-7}

All initial dry ingredients of the sugar-salt nutrient medium were investigated on SIMS mass-spectrometer. The same investigation was held with a part of the bacterial culture inoculum.

The cultures were grown in the thermostat at the optimal temperature 30°C during 48 hours. After completing each series, the obtained biological substance was separated using a centrifuge, cleaned in the distilled water and dried. The results of SIMS-analyses of cultures that were grown during experiments on isotope transmutation in regular and activated water are presented in Fig. 3.33.

3.7.5. *Conclusions*

The performed detailed studies have shown that activated water has a number of special or anomalous properties relative to similar, but nonactivated water.

It is worth to note the following essential features of activated water:

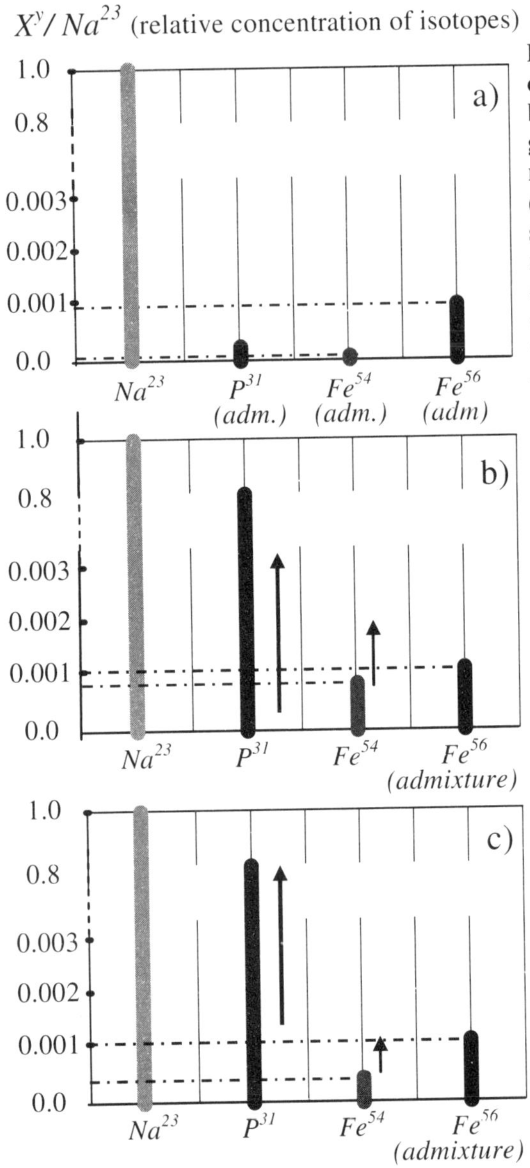

Fig. 3.33. Mass-spectrum of culture *Escherichia coli* (dried biological substances) that was grown in sugar-salt nutrient medium with presence of Na^{23} ($NaNO_3$ salt) and P^{31} (K_2HPO_4 salt) in activated and nonactivated water H_2O).

a) Control case - presence of Na^{23} (salt $NaNO_3$) and nonactivated water (P^{31} isotope was absent); $\eta(Fe^{54})/\eta(Fe^{56}) \approx 0,05$.
(adm) - admixture.

b) Experiment at transmutation conditions - presence of Na^{23} ($NaNO_3$ salt), P^{31} (K_2HPO_4 salt) and nonactivated water; $\eta(Fe^{54})/\eta(Fe^{56}) \approx 0,8$; Effectiveness of isotope transmutation is: $\lambda \approx N(Fe^{54})/N(P^{31}+Na^{23})\Delta t \approx 10^{-9}$ (synthesized Fe^{54} nuclei per s and per pair of P^{31} and Na^{23} nuclei).

c) Experiment at transmutation conditions - presence of Na^{23} ($NaNO_3$ salt), P^{31} (K_2HPO_4 salt) and activated water (t_{act} = 30 min); $\eta(Fe^{54})/\eta(Fe^{56}) \approx 0,4$; Effectiveness of isotope transmutation is: $\lambda \approx 0.5.10^{-9}$ (synthesized Fe^{54} nuclei per s and per pair of P^{31} and Na^{23} nuclei).

- It is characterized by a very great change (the reduction by 5 and more times) of the dielectric permittivity and the conductivity in the field of low and very low frequencies;
- A very significant reduction of both the viscosity and the coefficient of friction (the reduction by 10-100 times and more) is registered in the regions of small pressures acting on this water and of small speeds of its movement;
- The temporal nonmonotone behavior of the hydrogen index pH is revealed, and this change can be characterized by sharp maxima or minima for a certain type of activated water which are manifested in many days after the completion of the activation of water and have the form corresponding to phase transitions;
- Anomalous properties of activated water are preserved during a large time interval which can last many hours and days.
- Duration of preservation of the anomalous properties of activated water depends strongly on its temperature (it increases as the temperature drops and decreases as the temperature grows).
- It was observed for the first time that at low initial concentration of cells of investigated culture *Escherichia coli* in nutrient medium activated water have suppressed culture growth in 3 - 100 times!
- At the increasing of water activation time, inhibition of microbiological systems growth is more effective. In activated water the effect of anomalous division of grown cells takes place. In the control experiment, i.e. on the culture growth on the nonactivated agarized medium, the number of anomalous cells (4%) does not exceed the usual stochastic deviation.
- The presence of weak low-frequency magnetic fields in laboratory and environment leads to activation of water and to strong influences on the process of nuclear transmutation in growing biological systems (the effects of weakening and inhibition take place).

These properties can be used for the explanation of some effects of the anomalous influence of such water on biological objects and on phenomena at electrolysis and cold fusion in heavy water.

CHAPTER 4

Physical Aspects of Low Temperature Transmutation of Isotopes in Biological Systems

4.1. Low energy non-threshold nuclear reactions of synthesis with participation of charged particles

It is well known that the presence of the Coulomb barrier is the main obstacle to performing nuclear reactions of synthesis with low energy of interacting charged nuclei. In order to make such reaction possible, it is necessary that interacting particles be placed in the same spot simultaneously (within the range $R_n \approx 2.10^{-13}$ cm of action of nuclear forces). In this case, the cross section of nuclear reaction depends on the energy of reciprocal movement of nuclei and matches the "internal nucleus cross section"

$$\sigma_0(E) = \frac{S(E)}{E} \tag{4.1}$$

$S(E)$ is nuclear S-factor depending on a matrix element of the interaction energy of nuclei. $S(E)$ is the slowly changing function of energy, which with low relative energy of interacting particles in the case of non-resonance nuclear reactions, is constant $S(E) = S_0$ For example, $S_0 \approx 0.11 Mev\ bn$ for a non-resonance reaction of dd-fusion.

The kinetic energy E of reciprocal opposite movement of the nuclei, expressed in units of eV, leads to a situation when in the case of pure Coulomb interaction the presence of the Coulomb barrier

$$V(r) = \frac{Z_1 Z_2 e^2}{r}$$

determines the minimal distance of rapprochement of the nuclei, which is limited to

$$r_{\min} = \frac{Z_1 Z_2 e^2}{E} \approx \frac{14}{E(eV)} Angstroem \tag{4.2}$$

It can be seen that with low kinetic energy of the nuclei (on the level, corresponding to thermal energy E ≈ 0,025 eV) the nuclei can approach each other only by macroscopic distances, equal to

$$r_{\min} \approx 640 Z_1 Z_2 \ Angstroem \qquad (4.3)$$

The presence of electronic reflection causes faster decrease of the energy of interaction

$$V(r) = \frac{Z_1 Z_2 e^2}{r} e^{-r/a},$$

which allows particles with low energy to approach to the distance of $r_{\min} \approx a$, approximately equal to the range of reflection. For example, in the molecule of deuterium D_2 the balancing distance between the deuterons at the room temperature (kT = 0.025 eV) is equal 0.74 A. In order to achieve the same rapprochement of the nuclei in the absence of electronic reflection, the relative energy of the nuclei should be about 20 eV.

Meanwhile, with any reasonable density of electrons, the range of reflection exceeds the range of atomic forces action R_n by several orders.

An alternative to direct (classical) way of rapprochement of interacting nuclei is the quantum-mechanical tunnel penetration of one of the nuclei through a potential barrier created by another nucleus.

In such interaction, the cross section of the nuclear reaction

$$\sigma_0(E) = \frac{S(E)}{E} P(E) \qquad (4.4)$$

is determined by the product of "internal atomic cross section" $\sigma_0(E)$ and quasiclassical probability of penetration through the Coulomb barrier $V(r)$ (the "Gamov factor")

$$P = e^{-2W(E)},$$

$$W(E) = \int_{r_1}^{r_2} \sqrt{2\mu(\mathrm{V(r) - E})}\mathrm{dr}/\hbar = \sqrt{2\mu < V(E) >} \,|\, r_2 - r_1 \,|\, / \hbar \qquad (4.5)$$

In these expressions r_1 and r_2 are classical "turning points" of one particle's movement in the field of another; $\mu = m_1 m_2 / (m_1 + m_2)$ — reduced mass of interacting particles with masses m_1 and m_2;

$$< V(E) > = \left(\frac{1}{| r_2 - r_1 |} \int_{r_1}^{r_2} \sqrt{(\mathrm{V(r) - E})}\mathrm{dr} \right)^2$$ —average height of the potential barrier V(r), laying above the level of E.

It can be observed that the width of the barrier $r_1 - r_2$ (which is the linear function of the coordinate) affects the Gamov factor more significantly than its average height $<V(E)>$, included in (4.5) in the form of a square root.

In the case of pure Coulomb interaction $V(r) = Z_1Z_2e^2/r$ the value $W(E)$ is equal

$$W(E) = \frac{2\pi Z_1 Z_2 e^2}{\hbar v_0}, \tag{4.6}$$

where $v_0 = \sqrt{2E/\mu}$—asymptotical (at a large distance from the area of interaction) relative speed of nuclei motion.

For pure Coulomb interaction, probability density of finding nuclei in the same spot is determined by wave function (Davidov, 1963, in Russian), equal to

$$|\Psi_n(0)|^2 = \frac{2\pi Z_1 Z_2 e^2}{\hbar v_0^2} \exp\left(-\frac{2\pi Z_1 Z_2 e^2}{\hbar v_0}\right) \tag{4.7}$$

The rate of nuclear synthesis (probability of reaction per unit of time for one pair of interacting nuclei) is equal

$$\lambda_r = \Lambda |\Psi_n(0)|^2 \tag{4.8}$$

Here, $\Lambda = \frac{S_0 \hbar}{\pi \mu e^2}$— the constant of nuclear synthesis for a specific pair of nuclei, equal, for example, the value $\Lambda \approx 1{,}5.10^{-16}\ cm^3\ sec^{-1}$ for dd-interaction.

In normal conditions (i.e. without heating to a high temperature and without applying extra pressure), probability of the tunnel effect is very small.

Even for the optimal (with respect to the maximal transparency of the barrier) natural system—a molecule of deuterium D_2 with the distance between nuclei . $\Delta r \approx 0{,}74A$—probability of the tunnel effect at the room temperature is miniscule and equal to $P \approx 10^{-150}$. It results in a very slow rate of synthesis

$\lambda_r \approx 10^{-72}\ sec^{-1}$.

We would like to note, that all presented relations for probability of the tunnel effect have been obtained on the basis of the stationary Schredinger equation and, therefore, relate only to stationary interaction of the nuclei, although the process itself is never stationary. Non-stationary nature of any interaction is evident from the fact, that a system of interacting nuclei has its history and was formed at some point in the past.

Two obvious ways of increasing transparency of the barrier and achieving higher efficiency of synthesis follow from (4.5)—increasing energy or temperature of particles (this is the area of thermonuclear or accelerated synthesis) and reducing the width of the barrier at a low temperature (this is pyknonuclear synthesis with a compressed system, as well as muonic catalysis with reducing reflection of electrons by capturing a heavy μ-muon of at least one of the nuclei participating in the synthesis).

Pyknonuclear synthesis was proposed in the works of Zeldovich and Gernshtein (Zeldovich, 1960), as well as Harrison (Harrison, 1964). To achieve a practically plausible result, it is necessary to solve a very difficult problem of compressing condensed medium by 10^4–10^5 times, which can reduce the width of the barrier by 10–15 times.

If a particle is located in the dense (quasicrystalline) medium in a very deep parabolic well formed by adjacent nuclei, then its interaction with these nuclei positioned at a distance

$R_i \approx n_i^{-1/3}$ is ensured due to pyknonuclear reactions (Salpeter, 1969). The probability of a pyknonuclear reaction at each of the nuclei is given by the relations

$$\lambda \approx \sigma_0(E) < J_n > \approx \sigma_0(E)\frac{\omega_o}{\sqrt{8\pi^3}\sigma_{r0}}e^{-R_i^2/2\sigma_{r0}} = \sigma_0(E)\frac{\omega_o}{\sqrt{8\pi^3}\sigma_{r0}}e^{-1/2\sigma_{r0}n_i^{2/3}},$$

$$\sigma_0(E) = \frac{S(E)}{E},$$

$$< J_n > = <|v|> |\Psi(R_i)|^2,$$

$$|\Psi(R_i)|^2 = \frac{1}{(2\pi\sigma_{r0})^{3/2}}e^{-R_i^2/2\sigma_{r0}}, <|v|> \approx \omega_o\sigma_{r0}^{1/2}$$

Here, the three-dimensional wave function of a particle located in the parabolic well $\Psi(R_i)$ determines the probability $|\Psi(R_i)|^2 \equiv D$ of the localization of one nucleus at the place of another one; $\omega_0 = \sqrt{4\pi Z^2 e^2 n_i / M}$ is Debye frequency determining the energy of zero oscillations of a nucleus in the well; $E = \hbar\omega_0/2$; $\sigma_{r0} = \hbar/2M\omega_o$ - variance of the nucleus coordinate; $< J_n >$ is mean effective intensity of interacting particles; and $|v|$—mean velocity of an oscillating nucleus.

The pycnonuclear reactions are usually considered for astrophysical or very dense systems, and their probability increases with the concentration of nuclei n. For example, in liquid deuterium

($S_0 \approx 0.11 Mev\ bn$) with the density n = 10^{24} cm^{-3}, we have $\sigma_{r0}n^{2/3} \approx 2.5.10^{-3}$, and the probability of the reaction in an uncorrelated state is $\lambda_{dd} \approx 10^{-70}$ sec^{-1}. As the density increases to $n = 10^{27}$ cm^{-3} (in this case, $\sigma_{r0}n^{2/3} \approx 8.10^{-3}$), it grows to $\lambda_{dd} \approx 10^{-30}$ sec^{-1}; for $n = 10^{30}$ cm^{3} ($\sigma_{r0}n^{2/3} \approx 2.5.10^{-2}$), it reaches $\lambda_{dd} \approx 10^{-13}$ sec^{-1}.

Muonic catalysis was independently proposed by Frank (Frank, 1947) and Sakharov (1948) and later experimentally tested (Alvarcs, 1957). Complexity of such an attractive model was, first of all, due to the fact, that heavy μ-muons reflecting a nucleus' field are unstable particles having the life span of about 2.10^{-6} sec, which is insufficient for practical purposes.

The search for an optimal solution has resulted in the trend among all the leading world laboratories in the early 50's of the 20th century to choose thermonuclear mechanism of synthesis optimization as the main one, attempting to achieve success in quasi-stationary systems of prolonged sustenance of high-temperature, low-pressure plasma in magnetic traps, or using short-term compression by 100–1000 times and inertial holding at surrounding laser or beam compression of condensed system.

The experience of the last 50 years has shown, that this method of making nuclear reactions based on thermonuclear principles has proven to be extremely costly and ineffective in Earth conditions (in contrast to stars, where stabilization of hot thermonuclear plasma is naturally and efficiently achieved by forces of gravitation). The total expenses on thermonuclear programs have long ago exceeded $100 billion but the prospects of obtaining an energy efficient synthesis still remain just as uncertain now as they were in the beginning.

4.2. The phenomenon of "cold nuclear fusion"

There is one more branch of nuclear synthesis research, unrecognized by specialists in thermonuclear models, which can accommodate works, in which nuclear synthesis was observed in the absence of any high-energy influences.

The first reliably known works in this field were experiments on generating and registering helium from hydrogen by electrochemical means in the presence of palladium (Paneth, Peters, 1926). The general environment of skepticism and non-recognition of these experiments by leading nuclear physicists of the time resulted in dismissing these results as incorrect, which caused their relegation to the area of scientific curiosities. The authors of these works were chemists, not quite familiar with nuclear physics, which played a negative role in continuing research in this direction. They could not defend their experiments from criticism by "authorities" and prove their possibility.

In 1981-83 we have shown (Vysotskii, Kuzmin, 1981,1983) that in monocrystal targets like LiD the cross-section of fusion process with the participation of target nuclei (for example, deuterium planes or axis) and directed at Lindhard angle beam of moving nuclei (for example, tritium or deuterium nuclei) increases by 10-100 times relative to isotropic target. Such changes are based on the usage of specific channeling physics coherent phenomena, among them the main is so-called "overbarrier particles" or "overbarrier phenomena". In ordered crystal lattice exists very strong influence of internal electric field of crystal axes and planes on motion of moving charged particles and interaction between them and ordered crystal atoms and nuclei.

Equally little attention was given to later works, in which small emission of neutrons during mechanical or electromechanical influence on crystals containing deuterium (Kluev, 1986; Deriagin et al., 1986; Jaroslavskiy, 1986, 1989, in Russian) was observed. Perhaps, it is possible to explain the lack of public interest to these works by an important role played by ionic links in these crystals, which immediately suggested the effect of an accelerating mechanism due to appearance of powerful electrical fields at severing ionic links.

There was a surge of interest to these works after two groups of researchers from Utah (USA), who conducted independent experiments on electrolytic saturation of palladium and titanium, announced the results of their work.

Fleishmann and Pons from University of Salt Lake City (Fleishmann, Pons, 1998) announced that they had achieved a substantial emission of excess heat and registration of neutrons and tritium. Jones from BYU at Provo (Utah) also published his results, indicating same registration during the process of electrolysis of neutrons in the amount significantly higher that the background level (Jones, 1989). This effect was later called "cold nuclear fusion".

The results of an independent expertise of these studies were ambiguous.

Many laboratories around the world immediately corroborated the results obtained by Fleishmann, Pons and Jones, having discovered new, sometimes paradoxical (like dd-synthesis reactions without neutrons) features of performing efficient nuclear reactions with low energy, although they noticed, that the effect is not consistently duplicated.

At the same time, most largest laboratories, known for their activity in the field of thermonuclear research, made statements, in which they challenged the results of the experiments and showed their opposition to the very possibility of performing cold nuclear synthesis. Moreover, under their pressure, unprecedented administrative actions were taken aimed at discrediting such experiments and their banning. The methods of discrimination were akin to inquisition. In particular, an official ban on financing these experiments from government grants was issued in the United States.

Publication of experimental results in most prominent scientific publications was also prohibited. Science has not known such "witch hunts" since medieval times.

Nevertheless, the "genie was out of the bottle" and an attempt to stop scientific research in an alternative direction by administrative measures did not succeed. About 50 specialized scientific groups in Europe, Asia and North America were involved and continued their work on studying the nature of cold nuclear synthesis.

During the last decade, the technology and methodology of experiments on cold nuclear synthesis was considerably improved, more effective nuclear reactions have been found, detection techniques were altered, which led to major improvements in reliability and consistency of results. Part of these results is presented in the series of reviews (for example, Tzarev, 1990, 1992, 1993, in Russian).

Among the most convincing results, reliably registered generation of neutrons (Srinivasan, 1999; Arata, Chang, 1992; Takahashi, 1992, Mizuno, 2004), emission of excess heat (McKubre, 1990, 1991, 1992, 1994, 1995, 2000; Storm, 1991, 1996, 2000; Miles, 1990, 1993a,b, 1994, 1997; Miles, Bush, 1994; Miles et al, 2000; Arata, Chang, 1994, 1995, 1996, 1997, 2000; Takahashi, 1992; Patterson, 1994, 1997), transmutation of isotopes in a glowing charge in deuterium (Karabut, 1992, 1994; Savvatimova, 1992, 1998), and transmutation during electrolysis (Iwamura Y. et al., 2000) can be noted. Works on detecting transmutation of chemical elements in biological systems (Kervran, 1963, 1966, 1968, 1998; Komaki, 1965, 1967, 1993) are also directly related to this area, as well as the series of works on transmutation of both stable (Vysotskii, Kornilova, Samoylenko, 1996, 2000, 2001) and radioactive (Vysotskii, 2004) isotopes in microbiological cultures.

However, the main drawback of "early" cold nuclear synthesis—inadequate consistency of experimental results—has remained largely characteristic in even the most reliable works.

Other flaws in methods of registering cold nuclear synthesis (CNS) have been mentioned in several critical reviews (Morrison, 1994; Huitzenga, 1993; Taubes, 1993).

At present, such principal features of the phenomenon of cold nuclear synthesis, setting it apart from the area of medium and high energies as sporadic and non-stationary behavior in stationary conditions, almost complete absence of radioactivity of final products and very significant change of mass parts of different reaction channels, can be regarded as firmly established. These results are presented in hundreds articles and some monographies (e.g., Mallove, 1991; Mizuno, 1998; Kozima, 1998; Lindemann, 2001; Beaudette, 2002; Vysotskii, Kornilova, 2003; Krivit, 2004; Arata, 2008).

Meanwhile, the main and most important difference is, naturally, running the synthesis in conditions totally excluding any possibility of its occurring, based on theory, which successfully describes similar processes taking place with high energy.

The question arises: where might a logical error be contained—whether in extrapolation of the theory of nuclear reactions from the area of high energies to low energies, or the reason is that the theory explaining high energy reactions was constructed in such a way that it deliberately ignored some important patterns, peculiar to interactions at very low energy.

In our opinion, the second part of the question posed is closer to the truth and we will attempt to justify our viewpoint in the next section. Currently, there are over 400 works, in which—with a various degree of argumentation—different physical models are presented, capable, according to their authors, of explaining the phenomenon of "cold nuclear synthesis" or, at least, find ways to its explanation. It can be said, that the theory of cold nuclear synthesis is a particular mountaintop (like the famous Ferma theorem), which each esteemed theoretical physicist among those, who don't think that it is simply a mirage, feels obliged to conquer. The number of theoretical works on low energy nuclear fusion and their variety allow their sorting-out and classification. This process was initiated by the review article (Chechin et al., 1994) after the number of theoretical works had accumulated "critical mass". Among the recent reviews the work by Kirkinskiy and Novikov can be noted (Kirkinskiy, Novikov, 1998, in Russian). In these reviews, an approximately similar and logically coherent system of classification is used, which—with few exceptions—we are also using here. Latest works has updated the information in these reviews.

4.2.1 *Accelerating models*

(Vysotskii, Kuzmin, 1981,1983; Deriagin et al., 1986; Golubnichiy et al., 1989; Fleishman, Pons, 1989; Lipson, 1989; Cohen, Davis, 1989; Jouce, 1989; Gac et al., 1989; Cassandro et al., 1989; Goldanskiy, Dlidchik, 1989; Pool, 1989; Takeda, Takizuka, 1989; Tzarev, 1990, 1991, in Russian; Zelenskiy, Ribalko, 1990; Lin et al., 1990; Rabinovich, Worledge, 1990; Tabet, Tenenbaum, 1990; Jorne, 1990; Fillimonov, 1990; Zhang, 1991; Chekin, Tzarev, 1992; Yasui, 1992; Chekin et al., 1994; Kuhne, 1994; Varaksin et al., 1991; Fedorovich, 1991, 1992, 1993; Fedorovich, 1995, in Russian; Gonsales-Martin et al., 1993; Arata, Zang, 1995; Kazachkovskiy, 1995).

4.2.2 *Reducing the Coulomb barrier through reflection of electrons*

(Vaselli, 1989; Fujita, 1989; Burrows, 1989; Koonin, Nauenberg, 1989; Parnenter, Lamb, 1989; Mazitov, 1989, in Russian; Bhattacharjiee, 1989; Bussard, 1989; Rabinovich, Worledge, 1989; Karasevskiy, 1989, in Russian; Feng, 1989; Ichimanu, 1989, 1990, 1993; Jandel, 1990; Baljan et al., 1989; Davidov, 1989, in Russian; Rabinovitz, 1990; Azbel, 1990; Hora et al., 1990, 1993, 1998; Seeliger, 1991; Kim, 1991; Rambaut, 1992; Demidenko, Simakov, 1993, in Russian; Harrison, 1994; Fimin, 1996, in Russian; Violante, De Ninno, 1996).

4.2.3 *Coherent and interference effects involving lattice*

(Vysotskii , Kuzmin, 1981, in Russian; Turner, 1989; Bressani, 1989; Bush, 1990; Christos, 1990; Danos, 1990; Chatterjie, 1990; Preparata, 1990, 1991, 1995, 1996; Danos, Balyev, 1991; Vaidua, 1991; Kim, Zubarev, 1993; Kim, 1994; Kim et al., 1994; Fedorovich, 1994; Sapogin, 1994, 1997, in Russian; Li, 2006)

4.2.4 *Fermi and Bose nuclei condensation*

(Matsumoto, 1989; Vysotskii, Kuzmin, 1990, 1991, 1994; Greenland, 1990; Dienes, 1991; Mordlander et al., 1992; Vysotskii , Kuzmin, 1991; Vysotskii, 1994, in Russian; Chubb, Chubb, 1990, 1991, 1993; Waber, 1997; Kim, 2000)

4.2.5 *Reactions stimulated by lattice phonons*

(Schwinger, 1989, 1990; Petrillo, Sacchetti, 1989; Hagetstein, 1990, 1994, 1997, 1998; Baldo et al., 1990; Crawford, 1992; Tisenko, 1993, in Russian; Rambout, 1994; Rao, Chaptot, 1994; Liu, 1996; Swartz, 1996)

4.2.6 *Exotic chemistry*

(Mills, Farrel, 1990; Cerofolini, Re, 1990; Gann, Pokhodiashiy, 1990; Grysinsky, 1990; Barut, 1990; Moyges, 1991; Mills, Kneizys, 1991; Maly, Vavra, 1993; Samsonenko, 1996, 1997)

4.2.7 *Catalysis by heavy (transmuonic) particles*

(Shaw et al., 1989; Takahashi, 1989, 1991; Rafelski et al., 1990; Bazhutov, Vereshkov, 1992, 1993, in Russian; Nedospasov, Mudretskaya, 1995)

4.2.8 *Reactions under the influence of neutrons*

(Kim, 1989; Andennan, 1990; Hagelstein, 1990, 1991; Matzumoto, 1989, 1990a; Fisher, 1992; Lipson et al., 1993, in Russian; Lipson, Sakov, 1994, in Russian; Kozima, 1994, 1996)

4.2.9 *Peculiar nuclear effects*

(Schwinger, 1989; Becker, 1989; Ragheb, Miley, 1989; Paolo, 1989; Shibab-Elden et al., 1989; McVally, 1989; Kim, 1989, 1990; Frodi et al., 1990; Zakowitch, 1990; Kim et al., 1990; Rabinowitz, 1990a; Pokropivniy, Ogorodnikov, 1990, in Russian; Russel, 1990; Chatterjee, 1991; Kenny, 1991; Tsuchya, 1994; Mayer, Reitz, 1991; Mayer, Reitz, 1991; Yang, 1991; Istomin, Kaliev, 1994; Arata, Zhang, 1996; Takahashi, 1998; Dufour et al., 1998; Kim, Zubarev, 2000)

4.2.10. *Reactions under the influence of hypothetical magnetic monopoles*

(Urutskoev, 2002; Lochak,Urutskoev, 2004; Lochak, 2007; Adamenko, Vysotskii, 2006; Adamenko, Vysotskii, 2007).

4.3 Distinctive features of producing the "cold nuclear fusion" phenomenon in biological systems

4.3.1 *Is there a distinctive mechanism of producing low temperature nuclear synthesis in biological systems?*

Many different theoretical models of the phenomenon of "cold nuclear fusion" relate only to classical objects and physical models (nuclei, crystals, gas, plasma). Is there some peculiar nature of biological objects differentiating them from this natural series? Would it be correct to speak of "biological transmutation" or that is, really, only a regular transmutation, only occurring in biological systems? Which mechanism can sustain reactions of transmutation synthesis in growing microbiological cultures?

In order to answer these very complicated questions, we need, first of all, to verify the conceptual foundation of the processes, which could, in our opinion, sustain an efficient procedure of nuclear reactions at low temperatures and absence of other extreme conditions.

First, traditionally, nuclear physics (especially its part, studying the processes of interaction and diffusion) has been built upon the conviction, that since the process of paired interaction occurs in a very small area and is based on the effects of very powerful fields (internuclear and Coulomb), any influence of objects and fields, located far outside the area of interaction, is a priori impossible. In other words, with few exceptions, absolute isolation of atomic interaction from the outside influence was being postulated. The argument, that with higher energy and stronger impulse, wave properties of interacting particles (providing macro, as opposed to micro type of interaction) are less significant, was normally used as the justification. Moreover, it is believed that the very noticing, that the energy of interaction between nuclei is much higher than conventional energies of interaction in an environment, proves the appropriateness of ignoring the influence of the outside environment on nuclear processes.

In reality, due to wave properties natural to any micro object, this concept is by no means self-evident (not only in the field of low energies of interacting particles). This theory's inadequacy is manifested, for example, in the Mossbauer effect, which clearly demonstrates the dependency of processes of gamma-radiation interaction with nuclei from macroscopic properties of the matrix. The fact, that during interaction of fast electrons with a properly structured lattice in the course of channeling, wave properties of the particles are expressed better with higher energies (which is motivated by a proportional increase of the length of the area of coherent diffusion to the particles' impulse, so for relative particles it is significantly higher, than interatomic distance, and can reach macroscopic values), can serve as another illustration. Another corroborative factor is untrivial influence of dielectric permeability of thinned air on the dynamics of transforming cosmic rays of very high energy in the upper layers of atmosphere.

Second, practically all quantum-mechanical models, used for describing possible scenarios of running nuclear processes in problems related to low temperature synthesis, were based on the stationary (or quasi stationary) approach.

This opinion is supported by the mere fact that the concept of cross section of a reaction—fundamental for the analysis of synthesis processes—is used as a purely stationary characteristic, although, actually, cross section of a reaction (if understood as a probability of a reaction, associated with one target particle at a singular density of the stream of incoming particles) is, certainly, a non-stationary function of time, always equal to zero at the be ginning of the process. This conclusion follows directly from the Heisenberg expression for uncertainty correlation, because the uncertainty of energy of an incoming particle $\delta E \geq \hbar/2\delta t$ at the initial moment of interaction (with $\delta t \to 0$) is always infinitely large in relation to the area of conventional energies ΔE, where this channel of interaction is realized, which immediately proves small mass part of this particular channel of the reaction. To be fair, we need to note, that this effect of increasing δE is sometimes used for forecasting the above-barrier passage of particles.

Another mechanism, causing the non-stationary nature of the cross section of a reaction and its becoming zero at the beginning of an interaction process due to presence of relaxational processes, is examined in the book by Vysotskii and Kuzmin (Vysotskii, Kuzmin, 1989, in Russian). An adequate description of an interaction process in any mixed system (for example, in a system of two nuclei, which are always bound to a thermostat, made of atomic electrons, surrounding atoms, quantum, electromagnetic and peony fields, etc.) can be accomplished only on the basis of the matrix density method. This circumstance represents a major difference of the mixed system from the pure one consisting of two particles, which can be described with the help of two wave functions, if only idealized, purely paired interaction is considered, without regard to any interaction with the environment. Analysis of these problems shows, that the cross section of a reaction is always a non-stationary function of time, which is equal to zero or approaching zero for periods of time, that are much shorter than the time of relaxation of non-diagonal matrix elements, and changes to its stationary condition only for a period of time, significantly greater than relaxation time (Vysotskii, Vorontzov, 1974, in Russian). And although the time of relaxation of non-diagonal matrix elements declines with weaker connection to the thermostat, it is never equal to zero. Hence a simple logical conclusion—for very short periods of time after "activating interaction", the system's response to this interaction is zero. Naturally, the same applies to such a routine result of interaction as diffusion in the Coulomb field.

These phenomena are related to the analysis of patterns of the cross section of a reaction (diffusion, in particular) in a short time since the beginning of interaction. However, the non-stationary nature of the reaction (including turning the cross section of diffusion into zero) will be also manifested in later periods of time.

In our opinion, accounting for these circumstances is one of specific keys to this problem, while an adequate theory should definitely take them into account.

Apparently, it can be applied primarily to growing microbiological systems.

These systems—thanks to continuity of life functions—are non-stationary by definition and naturally combine relative stability and absolute changeability (especially on the molecular level) and this sets them apart from stationary crystals or quasi-stationary plasma. Meanwhile, the very character of the processes in transporting and enzymatic systems, with continuous protein synthesis and similarly permanent recreation of cells, is based directly on molecular processes given direct impact and control (in figurative, functional, as well as the immediate spatially-conforming sense) of molecular complexes of different kinds.

To support this thesis, it would be enough to give an example of selective transportation in membrane channels with constantly changing diameter and shape, reacting adaptively and pointedly to an ion type. There are no known analogues of such combination of changeability and stability in the inanimate nature.

These circumstances clearly demonstrate the necessity of consistent accounting for quantum-mechanical characteristics of interacting particles (in this case, nuclei, participating in synthesis and transmutation), as well as other objects, immediately bordering the area of interaction.

We believe that manifestation of these characteristics represents the distinctive feature of making low temperature nuclear synthesis and transmutation in biological systems. Thus, although the term "biological transmutation of isotopes" is technically incorrect and should be replaced by a more appropriate definition of "transmutation occurring in biological systems", the specific character of transmutation in dynamic by their nature biological systems definitely exists, while it makes the process of transmutation itself much more efficient.

4.3.2 *Potential mechanism of non-barrier nuclear synthesis in optimal potential quantum wells in the non-stationary mode*

Which model can accommodate all the above requirements to the greatest extent?

We think that the most effective in the case of dynamic biological systems is the mechanism, which we have suggested earlier (Vysotskii, 1994; Vysotskii et al., 1996, 2000), capable of removing—for a brief time—the influence of the Coulomb barrier of a nuclear reaction occurring in a potential well with a structure close to parabolic.

Let's examine this mechanism more closely using the example of the synthesis reaction of the Mossbauer isotope $Mn^{55} + d^2 = Fe^{57}$ in growing microbiological systems, which we conducted earlier (Vysotskii, Kornilova, Samoylenko, 1996) and discussed above.

Let's assume, that an atom of manganese is located in the middle of a three-dimensional potential parabolic well. Because of the process of dissociation, there is a large quantity of free deuterons in a liquid environment (at $T = 300K$, probability of dissociation is equal $\xi \approx 10^{-10}$). When a deuteron d falls into the well in the result of dissociation, combination Mn + d is formed, $V(r) = Ze^2/r > 0$ при $r < a/Z$ ($a \equiv \hbar^2/me^2$) which is further augmented by a collectivized electron from the lattice. This combination is not capable of forming a molecule in the free space from covalent or ionic links, but it can create a stable quasi molecule (MnD)+ at very low temperatures due to Van-Der-Vaals forces, for example. In a quantum system situation is more complicated.

It is known, that the process of diffusion for charged particles is determined, mainly, by Coulomb repulsion of nuclei and $V(r) = Ze^2/r > 0$ at $r < a/Z$ ($a \equiv \hbar^2/me^2$).

Can this energy be merely a small correction, not affecting the pattern of nuclei movement in a quantum system? If so, then under what conditions is it possible?

In order to answer these questions, let's consider the process of interaction and diffusion of particles in a well.

We will examine the problem of diffusion of a particle (deuteron) on a manganese nucleus in the impulse presentation, which, in this case, better describes the problem.

In order to investigate the essence of the process of short-term exclusion of the Coulomb barrier's influence, we begin with considering a modeling problem, when one of the interacting particles is permanently located in the well and is motionless (the classical center of diffusion), and another one moves according to the laws of quantum mechanics, accounting for interaction with the field of the well and the other particle.

The wave function of a particle moving in a potential well $U(r)$ has the form of coherent superposition

$$\psi(r,t) = \sum_p C_p\psi_p(r)\exp(-iE_pt/\hbar), \tag{4.9}$$

determined by the specter of energy levels E_p of a particle in the well and its own coordinate functions of a particle $\psi_p(r)$, corresponding to these levels of energy.

The Schrodinger equation can be written as follows

$$i\hbar\partial C_{p_0}/\partial t = \sum_p C_p V_{p_0 p}\exp[i(E_p - E_{p_0})/\hbar,$$

$$V_{p_0 p} = \int \Psi^*{}_{p_0}(r)V(r)\Psi_p(r)d^3r \tag{4.10}$$

Let's make the problem more precise. Let's assume, that the particle fell into a well at the moment of time $t = 0$. At the initial point of time, the particle's condition is characterized by energy E_{p_0} and the wave function $\Psi_{p_0}(r)$ (provided that $C_{p_0} = 1$).

In the Borne approximation, the process of diffusion of a particle is defined by the equation

$$i\hbar\partial C_p/\partial t = C_{p_0}V_{pp_0}\exp[i(E_{p_0} - E_p)/\hbar] \tag{4.11}$$

The question of whether using the Borne approximation in the problem of diffusion on the Coulomb potential is justified was rigorously discussed by Bohm in his famous book on quantum theory (Bohm, 1965, in Russian):

"The Borne approximation often yields remarkably good results, despite the fact, that none of the general criteria for its applicability is fulfilled. The following reasons give explanation to that.

The Coulomb force cannot be achieved near an atom's border, however, it is small, because electrons reflect a major part of the nucleus charge. Therefore, the Borne approximation can be used in this case because V(r) is small. Inside the atom, the potential of V(r) is so great, that we can expect that the Borne approximation will be completely useless, but since reflection and the Coulomb potential are not present here, a precise result will approach the result obtained using the Borne approximation.

Therefore, the Borne approximation gives a comparatively good result for the whole atom, which, however, cannot be substantiated by the general conditions."

Solving the equation of movement with the initial condition $C_{p_0}(0)=1$, $C_p(0)=0$ produces

$$|C_p(t)|^2 \equiv W_{p_0 p}(t) = |V_{p_0 p}|^2 \frac{\sin^2[(E_{p_0}-E_p)t/2\hbar)}{[(E_{p_0}-E_p)/2]^2} \qquad (4.12)$$

This expression holds for the probability of diffusion of a particle with transformation from the initial state $\Psi_{p_0}(r)$ with energy E_{p_0} into an arbitrary different state $\Psi_p(r)$ with energy E_p due to interaction with the diffusing center (another particle).

Notably, this probability is a complicated non-monotonous function of time.

Further analysis of the problem is principally different in two cases:

(a) for the traditionally used model of paired interaction in free space and without considering influences from other objects and fields, mentioned above;

(b) for the problem of interaction of particles in a real quantum field $U(r)$, being studied by us.

(a) We shall first consider the problem of diffusion based on the traditional model of diffusion in free space.

In the case of free space, approximated by an infinitely increasing volume of quantization V_0, we can introduce a three-dimensional density of final conditions in a pulsed space, corresponding to different pulse values

$$\rho(p) \equiv dN/dp = (d/dp) \iiint d^3x \iiint d^3p/(2\pi\hbar)^3 = V_0 p^2/(2\pi^2\hbar^3) \qquad (4.13)$$

We can see that the distance between neighboring resolution states in the pulsed space with different pulse values

$$\rho(p)^{-1} = (2\pi^2\hbar^3)/V_0\, p^2 \qquad (4.14)$$

with an infinite increase of dimensions of the area of quantization ($V_0 \to \propto$) decreases infinitely.

After that, the differential probability of diffusion of a particle in the direction of a spatial angle δΩ and the final impulse p in the interval *dp* assumes the form

$$dW(p, t) = \rho(p) \mid C_p(t) \mid^2 dp_x dp_y dp_z = \rho(p) \mid C_p(t) \mid^2 p^2 dp\delta\Omega \qquad (4.15)$$

Then, using the non-relativist link of the impulse energy $Ep = p^2/2M$, we can introduce the density of final conditions of the energy

$$\rho(E_p) = \sqrt{2M^{3/2}E_p}\ \rho(p)$$

and bring (4.15) to the form, determining the differential probability of diffusion of a particle in the same direction ... with a simultaneous transformation into a state with energy E_p in the interval dE_p

$$dW(E_p, t) = \rho(E_p) \mid C_p(t) \mid^2 dE_p\ \delta\Omega \qquad (4.16)$$

After substituting (4.13), we obtain expressions for full probability of diffusion of a particle in the direction δΩ at the time *t*

$$\delta W(t) = \rho(E_{p_0})\delta\Omega \int \mid V_{p_0 p} \mid^2 \frac{\sin^2[(E_{p_0} - E_p)t/2\hbar)}{[(E_{p_0} - E_p)/2]^2} dE_p \qquad (4.17)$$

The sequential procedure is the following.

Using one of the definitions of the Dirac delta-function

$$\frac{\sin^2 xt}{\pi t x^2} = \delta(x), \qquad (4.18)$$

occurring when t is large, (4.17) is easily integrated, which produces the final well known expression for full probability of diffusion into the spatial angle δΩ

$$\delta W(t) = (2\pi/\hbar)\ t\ \rho(E) \mid V_{p_0 p} \mid^2\ \delta\Omega \qquad (4.19)$$

It follows from (4.19)—after substituting in the explicit expression for $\rho(E_p)$ —that we can find the formula for the stationary cross section of diffusion (Bohm, 1965)

$$\sigma = \mid V_{p_0 p} \mid^2 (2\pi M/\hbar^2)^2 \qquad (4.20)$$

From (4.18) we can see that the probability of diffusion increases monotonously with increasing time, without becoming zero in any point. The cross section of a particle diffusion (4.19) at the center of diffusion is also different from zero, which, in the end, corresponds to a strong diffusion and a small probability of one nucleus getting to the area of localization of the other.

These formulas are normally used as universal (including the analysis of conditions for producing nuclear reactions), although they were obtained with some limitations (in particular, in applying the condition of the infinite increase of quantization volume *V*0, related to transformation towards free space and ignoring other objects and in using the asymptotical formula (4.18), which holds true only at very large values of *t*).

(b) We shall now perform an alternative analysis of the problem about particles interaction and diffusion in a concrete quantum field *U*(*r*) with discrete distribution of levels.

Let's consider the situation, which refers to diffusion in a three-dimensional parabolic well, the potential energy of which has the form

$$U(r) = kr^2/2 = M\omega_0^2 r^2/2 \qquad (4.21)$$

Here, $\omega_0 = \sqrt{2U_0/MR^2}$, $k = M\omega_0^2$, $U_0 \equiv U(R)$ —the height of the walls of the potential parabolic well.

The spectum of energy levels for a potential parabolic well is also equidistant and is measured from the bottom of the well by the formula

$$E_n = \hbar\omega_0 (n+3/2),\ n = 0, \pm 1, \pm 2, \ldots, \qquad (4.22)$$

The calculated above probability of diffusion with occupation of a specific level of energy E_p is determined by the solution (4.13)

$$W_{p_0 p}(t) \equiv |V_{p_0 p}|^2 F((E_{p_0} - E_p), t),$$

where relation to time and energy is determined by the function

$$F(E,t) = \frac{\sin^2[(E_{p_0} - E_p)t/2\hbar)}{[(E_{p_0} - E_p)/2]^2} \qquad (4.23)$$

After substituting energy for the levels E_n (4.22) into the function (4.23) we obtain

$$F(E,t) = \frac{\sin^2[\omega_0 t(n_0 - n)/2)}{\hbar^2[\omega_0(n_0 - n)/2]^2} \qquad (4.24)$$

It is evident, that the function *F*(*E*, *t*) and the probability of diffusion related to it are oscillating functions of time, characterized by equidistant distribution of zeros.

That means, that for the time series (beginning at the moment of falling of a particle into a well at $t = 0$)

$$\tau_k = 2\pi k/\omega_0,\ k= 0, \pm 1, \pm 2, \ldots, \qquad (4.25)$$

in a specific potential parabolic well, the probability of diffusion of one particle into another with transformation of the diffused particle from the initial level $E_{n_0} \equiv E_{p_o}$ to any of the levels $E_n \equiv E_p$ turns to zero.

This situation is shown on Fig. 4.1 for three consecutive moments of time.

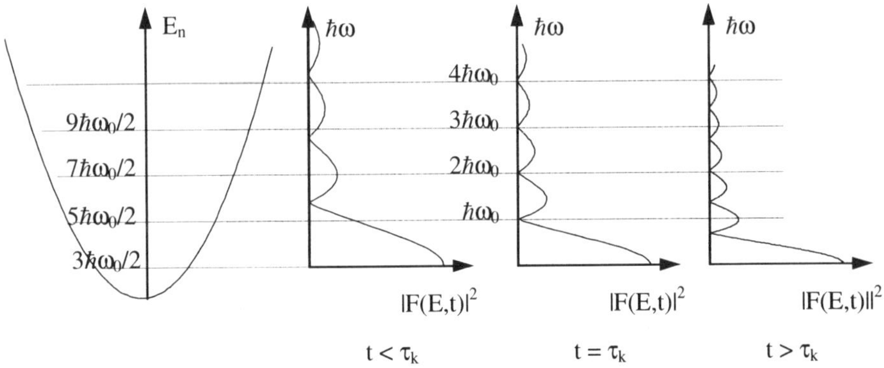

Fig. 4.1. Correlation between energy levels E_n in a three-dimensional potential parabolic well and full probability of diffusion, for the case of short ($t < \tau_k$), optimal ($t = \tau_k$), and large ($t > \tau_k$) mutual time of staying of two interacting particles in this well

An apparent reason for such an effect is that at the optimal moments of time τ_k there are no components in the spectrum of interaction between particles with energy, corresponding to transformation between levels or diffusion in the direction of any allowed state of itself. There is an interference suppression of the effect of Coulomb diffusion. And since there are no levels and conditions that are different from self-conditions in a quantum field $U(r)$, it means, that in these moments diffusion is prohibited. It means, in turn, that in these moments of time τ_k (and, naturally, within small time intervals close to all values of τ_k) the effect of Coulomb repulsion is not significant. In these moments of time, particles' states do not depend on their interaction, but are determined exclusively by the properties of quantum field.

It is possible to conduct a generalization of the case being studied, when one of the particles is fixed in a well, while another particle is moving according to the character of its interaction with the first particle and the field of a potential well, to the case, when both particles are moving.

If the particles are identical, they have a common system of quantum levels.

It means, that preventing one particle from diffusion at the moment of time τ_k (see 4.25) automatically prevents diffusion of another particle.

If the particles are different, each of them is characterized by its own system of energy levels in a potential well, its peculiar frequencies $\omega_0^{(i)}$ and, accordingly, its values of time $\tau_k^{(i)}$ (see 4.25), when diffusion is impossible. However, since the only reason of one particle's diffusion is its interaction with the other particle, the impossibility of diffusion at certain moments of time τ_k for one particle corresponds to the similar impossibility of diffusion for another one.

Therefore, the effect of Coulomb barrier is removed at those points of time, which match sets of values $\tau_k^{(i)}$ for each of the particles.

At these moments of time, the Gamov factor $P = e^{-2W(E)}$ becomes zero and the effective cross section of a synthesis reaction $\sigma = (S(E)/E)\ P(E)$ corresponds to the internal nuclear cross section $\sigma_0 = (S_0/E)$.

The fact, that at points of time τ_k localization of particles in the same well is determined not by their interaction, but by an independent influence of a potential quantum field, means, that the wave function of one particle at the location of another $\Psi_n(0)$ does not depend on the tunnel effect and, thus, may be arbitrary (it is determined solely by the conditions of quantomizing in the field U_0).

Specifically, in a quasi-classical approximation of wave functions $\Psi_n(\vec{r}_A), \Psi_\alpha(\vec{r}_B)$ of each of the particles is determined by the expressions

$$|\Psi_n(\vec{r}_A)|^2 = \frac{2}{\pi^2 R_n^3 \sqrt{1-(r_A/R_n)^2}}, r_A \le R_n;$$

$$|\Psi_\alpha(\vec{r}_B)|^2 = \frac{2}{\pi^2 R_\alpha^3 \sqrt{1-(r_B/R_\alpha)^2}}, r_B \le R_\alpha$$

Here, $R_n = \sqrt{2E_n/M\omega_0^2} = \sqrt{2\hbar(n+3/2)/M\omega_0}$ —classical amplitude of oscillations of a particle with energy E_n in a potential parabolic field.

The reaction speed in such a delocalized distribution of particles participating in the reaction is described by the formula

$$\lambda_r = \Lambda \int |\Psi_n(\vec{r})|^2 |\Psi_\alpha(\vec{r})|^2\, d^3r \approx \frac{16}{\pi^2 R_n^3 R_\alpha^3} \int_0^{R_{max}} \frac{r^2 dr}{\sqrt{1-(r/R_n)^2}\sqrt{1-(r/R_\alpha)^2}} \quad (4.26)$$

which is a generalized form of (4.8).

In (4.26), R_{max} is determined by relations

$R_{max}=R_n$ if $R_n \leq R_\alpha$ and $R_{max}=R_\alpha$ if $R_n \geq R_\alpha$.

If $R_n >> R_\alpha$ or $R_\alpha >> R_n$, from (4.26) we obtain

$$\lambda_r \approx \Lambda|\Psi_{n,\alpha}(0)|^2 = 2\Lambda/\pi^2 R_{max}^{\ 3} \qquad (4.27)$$

In particular, when $R_{max} = 3$ A, the speed of the reaction for dd-synthesis is equal $\lambda_r \approx 10^6$ sec^{-1}, which is 10^{78} times faster, than dd-synthesis reaction in the molecule of D_2

There is another important factor.

It follows from (4.26), that the optimization condition for nuclear synthesis requires using small volumes V_0 of quantomizing potential wells and big values of Λ. The question of the optimal value V_0 will be discussed later. Regarding Λ the following needs to be mentioned.

The nuclear synthesis constant for a particular pair of nuclei $\Lambda= S_0\hbar/\pi\mu e^2$ is determined by the value of the nuclear parameter S_0 and available mass of interacting particles μ. Parameter S_0 depends on the type of interacting nuclei and increases drastically with higher atomic number.

For example, for interaction of two protons (a pp-fusion reaction) $S_0 \approx 3{,}5.10^{-25}$ MeV•bn, which corresponds to the minimal value of $\Lambda_{pp} \approx 10^{-39}$ cm^3sec^{-1} and a very small speed of the synthesis reaction.

For interaction of a proton with a deuteron (pd-synthesis) $S_0 \approx 2{,}6.10^{-5}$, which corresponds to $\Lambda_{pd} \approx 5.10^{-20}$ cm^3•sec^{-1}.

For dd-interaction, $S_0 \approx 0.11$ MeV•bn and $\Lambda_{dd} \approx 1{,}5.10^{-16}$ cm^3•sec^{-1}.

For interaction of carbon nuclei (C^{12}-C^{12}-synthesis), the nuclei parameter $S_0 \approx 8{,}8.10^{16}$ MeV•bn (Ishimaru, 1999), which gives $\Lambda_{C^{12}C^{12}} \approx$ 20 sm^3•sec^{-1}.

For C^{12}-O^{16} interaction of nuclei $S_0 \approx 1{,}15.10^{21}$ MeV•bn (Ishimaru, 1999), which corresponds to $\Lambda_{C^{12}O^{16}} \approx 2{,}3.10^5$ cm^3•sec^{-1}.

Finally, for interaction of oxygen nuclei (O^{16}-O^{16}-synthesis), the nuclear parameter still increases ($S_0 \approx 2{,}3.10^{28}$ MeV•bn (Ishimaru, 1999)), which gives a very large constant of nuclear synthesis $\Lambda_{O^{16}O^{16}} \approx 3{,}9.10^{12}$ cm^3•sec^{-1}. Such a big value of $\Lambda_{O^{16}O^{16}}$ causes a very fast speed of transmutation.

We have all the reasons to believe, that such a rapid increase of S_0 and . with higher atomic number will be characteristic for all nuclei, which participate in an energy-efficient synthesis.

This situation leads to a new surprising conclusion about a significant increase of the probability of a barrier-less synthesis with increasing atomic number of the nuclei, participating in the reaction!

Qualitatively, such an effect is similar to preventing Coulomb diffusion of electrons moving in a conductor on ions, inside a filled-up Fermi sphere.

Nevertheless, in the case of a Fermi sphere, prevention of diffusion is related to the lack of free (unoccupied) states on different energy levels, while, in our case, such prohibition is attributed to the absence of the necessary specter components of a diffusing potential at certain moments of time (components, matching the frequencies of all possible transformation between energy levels in a well).

Obviously, this effect is directly relevant to the notion of non-stationarity of the diffusion process (some moment of time, when interaction has begun), as well as the influence of a potential well's structure on diffusion process.

It follows directly from the data obtained during performance of boundary transformations.

Clearly, with an unlimited increase of time elapsed since the beginning of the process of diffusion (with $t \to \infty$)), the rapidly-oscillating function $\sin^2[\omega_0 t(n_0 - n)/2)]$ in expression (5.44) becomes equal *1/2* while expression (4.24) assumes its always different from zero asymptotical form

$$F(E, t \to \infty) = \frac{1}{2\hbar^2[\omega_0(n_0 - n)/2]^2} \neq 0. \qquad (4.28)$$

Naturally, the probability of diffusion –

$$W_{p_0 p}(t) \equiv |V_{p_0 p}|^2 F((E_{p_0} - E_p), t)$$

or, ultimately, the effect of Coulomb repulsion—will be different from zero for transfers to any energy levels.

Similarly, if the size of a quantum well becomes infinitely big (provided, that $R \to \infty$, $\omega_0 = \sqrt{2U_0/MR^2} \to 0$), optimal moments of suppression of diffusion $\tau_k = 2\pi k/\omega_0 \to \infty$ will also become unattainable.

Moreover, with such an unlimited increase of the size of a quantum system, because of elimination of discreet levels

$$E_{n+1} - E_n = \hbar\omega_0 \to 0$$

we have to switch to using continuous approximation, which has been discussed earlier.

For this case also, full suppression of diffusion to all levels at once, is impossible.

Discovered features of brief suppression of the Coulomb barrier influence put certain requirements on system features (most of all, to its temperature).

A particle falls into a well and can leave it as a result of a thermally stimulated diffusion. Time, spent by a particle in a well

$$t=(2\pi/\omega_0)\exp(\varphi_i/kT)$$

is determined by temperature T and an energy interval to the well's top φ_i (energy of activation). At high T, the value of t is very small ($t < \tau_k$), and a particle leaves the well in much less time, than is required for suppression of the Coulomb barrier. At a very low temperature, a particle can stay in the well longer, than is needed for satisfying the condition $t > \tau_k$, which leads to possibility of suppression of the Coulomb barrier of a reaction at a moment of time $t = \tau_k$.

Thus, there is a limit on temperature of the environment $T < T_{cr}$, where

$$k_B T_{cr} = \varphi_i \, ln(\omega_0 \tau_k / 2\pi) = \varphi_i \, ln(k) \qquad (4.29)$$

Considering, that $ln(k)$ always satisfies the condition $1 < ln(k) < 10$, the environment temperature criteria is met if it's lower than activation energy of the well.

Let us make one more remark.

In order for both particles to spend a long time staying together in a common potential well (and do not "notice" each other by moving completely independently of the other particle if the above condition is fulfilled) it is necessary, that the average energy of interaction of both nuclei, taking outside influence into consideration, should be equal to zero (speaking in formal quantum mechanics terms, diagonal matrix energy elements of interaction of particle n and particle α should be equal to zero).

$$V_{n\alpha,n\alpha} = \int \Psi_n^*(\vec{r}_A)\Psi_\alpha^*(\vec{r}_B)\hat{V}(|\vec{r}_A - \vec{r}_B|)\Psi_n(\vec{r}_A)\Psi_\alpha(\vec{r}_B)d^3r_A d^3r_B \equiv$$
$$\int |\Psi_n(\vec{r}_A)|^2 |\Psi_\alpha(\vec{r}_B)|^2 \hat{V}(|\vec{r}_A - \vec{r}_B|)d^3r_A d^3r_B \approx 0 \qquad (4.30)$$

In quasi-classical approximation, this condition assumes the form

$$V_{n\alpha,n\alpha}=\frac{32}{\pi^2 R_n^3 R_\alpha}\left\{\int_0^{R_\alpha}\frac{dr_A}{\sqrt{1-(r_A/R_\alpha)^2}}\left[\int_0^{R_\alpha-r_A}\left(\sqrt{1-\left(\frac{r_A-r}{R_\alpha}\right)^2}-\sqrt{1-\left(\frac{r_A+r}{R_\alpha}\right)^2}\right)V(r)r_A r dr+\int_{R_\alpha-r_A}^{R_\alpha+r_A}\sqrt{1-\left(\frac{r_A-r}{R_\alpha}\right)^2}V(r)rdr\right]+\int_{R_\alpha}^{R_n}\frac{dr_A}{\sqrt{1-(r_A/R_n)^2}}\int_{r_A-R_\alpha}^{r_A+R_\alpha}\sqrt{1-\left(\frac{r_A-r}{R_\alpha}\right)^2}V(r)r_A r dr\right\}\approx 0$$

(4.31)

in the case $R_n \geq R_\alpha$ (if $E_n \geq E_\alpha$) and

$$V_{n\alpha,n\alpha}=\frac{32}{\pi^2 R_\alpha^3 R_n}\left\{\int_0^{R_n}\frac{dr}{\sqrt{1-(r/R_\alpha)^2}}\left[\int_0^{R_\alpha-r}\left(\sqrt{1-\left(\frac{r-r_A}{R_n}\right)^2}-\sqrt{1-\left(\frac{r+r_A}{R_n}\right)^2}\right)V(r_A)r\,r_A dr_A+\int_{R_n-r}^{R_n+r}\sqrt{1-\left(\frac{r-r_A}{R_n}\right)^2}V(r_A)r_A dr_A\right]+\int_{R_n}^{R_\alpha}\frac{dr}{\sqrt{1-(r/R_\alpha)^2}}\int_{r-R_n}^{r+R_n}\sqrt{1-\left(\frac{r-r_A}{R_n}\right)^2}V(r_A)r_A dr_A\right\}\approx 0$$

(4.32)

in the case, when $R_n \leq R_\alpha$ (if $E_n \geq E_\alpha$).

Familiarity with the actual form of particle interaction energy is necessary for an analysis of the possibility of fulfilling conditions (4.31) and (4.32).

The actual form of this energy is well known when we deal with protons or deuterons (Davidov, 1963), as well as the case of a system of two protons and electrons

$$V(r)=\frac{e^2}{a\rho}\left(1+\frac{5}{8}\rho-\frac{3}{4}\rho^2-\frac{1}{6}\rho^3\right)e^{-2\rho},\ \rho=\frac{r}{a},\ a=\frac{\hbar^2}{m_e e^2}$$

The diagram of this function is presented on Fig. 4.2.

Please note, that the interaction energy of two protons in the presence of two electrons is a variable polarity function of the coordinate.

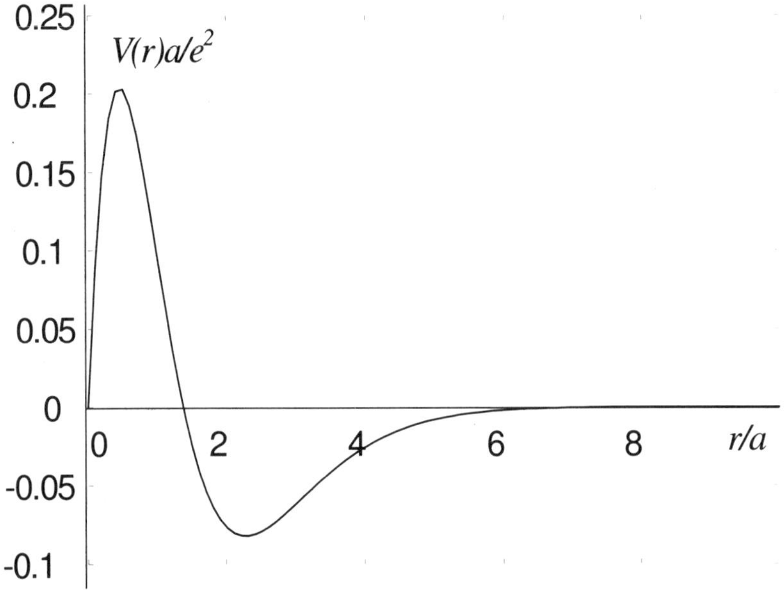

Fig. 4.2. Energy of interaction of two deuterons in a d-e-e-d system.

Interaction energy of two heavier nuclei *V*(*r*) in an environment containing electrons is also a variable polarity function of distance between nuclei (see Fig. 4.3).

For large distances $r \geq a \equiv \hbar^2/me^2$ this energy is actually determined by parameters of ionic links and characteristics of Van-Der-Vaalse forces of the system Mn^+–e–d, provided that $V(r) < 0$.

In the area of intermediate distances $a/Z < r < a$, this energy can be characterized by Thomas-Fermi approximation

$$V(r) = \{Ze^2/r\}\ \chi(rZ^{1/3}/0{,}885a),$$

with possible values $V(r) < 0$ and $V(r) > 0$ in this interval.

In the area of small distances $r < a/Z$, atomic interaction corresponds to Coulomb repulsion of the nuclei and $V(r) = Ze2/r > 0$.

On Fig. 4.3, the spatial structure of interaction energy between nuclei Mn and d (with the presence of compensating electrons), is presented.

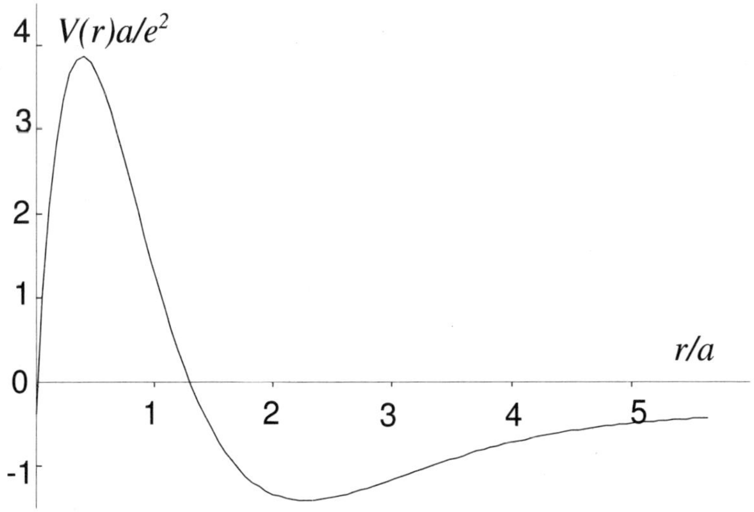

Fig. 4.3. Expression structure for energy of interaction between nuclei Mn and d with the presence of electrons.

At first sight, it appears, that since Coulomb interaction is much stronger than other types of interaction, fulfilling the condition $V_{n\alpha,n\alpha} \approx 0$ can be achieved only in a very large well with $R_\alpha, R_n \to \infty$..

Obviously, such a solution does not provide optimization of rate and probability of synthesis (4.26), (4.27), because then

$\lambda_r \approx \Lambda|\Psi_{n,\alpha}(0)|^2 = 2\Lambda/\pi^2 R_{max}{}^3 \to 0$.

A more rigorous examination reveals, that fulfilling the condition $V_{n\alpha,n\alpha} \approx 0$ is also possible in a microscopic well. It is based on the fact, that in the expression for $V_{n\alpha,n\alpha}$ potential energy is always part of combinations with a squared coordinate (resulting from integration in a spherical system of coordinates), which means that integral contributions of areas, where $V(r) > 0$ and where $V(r) < 0$, are quite compatible (see Figs. 4.2 and 4.3).

In result, it is almost always possible to find such an optimal size of a well, that these contributions would cancel out, thus providing a sought result $V_{n\alpha,n\alpha} \approx 0$.

4.3.3 *Possible scenario of running barrier-less nuclear synthesis in growing biological systems.*

Combination of obtained conditions for realization of the effect of brief removal of the Coulomb barrier influence on movement of interacting nuclei and efficiency of nuclear reactions can be provided in the volume of optimal structural elements of the surrounding matrix (for example, in spherical micro-cavities with diameter $2R_0 \approx 10 \div 12A$, micro cracks $4 \div 5A$ in width and other micro distortions, creating two- or three-dimensional potential wells (Vysotskii, Kuzmin, 1990)).

Fulfilling initial conditions for the parameters of potential wells of a catalytic matrix allows creating a "barrier-less" low temperature (non-accelerating and non-tunnel!) nuclear synthesis. Please note, that these conditions are quite stringent.

The size (radius) of potential wells R_0 and slope of the relief U_{max}/R_0^2 for example, ought to be precisely determined.

According to our estimates, for a d–e–d system (Fig. 4.2) the optimal would be a well with $R_0 \approx 2A$. For a system Mn–d (Fig. 4.3), the well size of $R_0 \approx 4{,}5A$ would be optimal. It can be seen that the optimal size of these wells is compatible with inter-atomic distance. Because of that, such wells would not be closed in broadly interpreted terms (their "walls" would not consist of atoms). Most likely, such wells would emerge in the central area of axiallysymmetrical wells of large size.

Meanwhile, it is very difficult to calculate these parameters precisely due to many factors, resulting from distortions of lattice. It is very unlikely, that structure and size of micro distortions will, by chance, precisely match the requirements (quadratic profile and optimal slope of potential energy, optimal size).

This may be the reason for low efficiency and poor duplication rate of experiments on CNS in fixed solid objects. To optimize CNS process, an "adjusting" mechanism for existing parameters of micro distortions, bringing them to optimal, is needed.

Growing microbiological cultures provide good conditions for producing CNS. There is rapid structural transformation in the area of culture's growth.

It is accompanied by continuous generation of structural micro distortions with altering dimensions and potential profile with time. Adaptive changes of diameter and shape of membrane channels take place. Location of enzymes and proteins changes in relation to each other (for example, during the DNA replication process and protein synthesis near a DNA polymerase and other enzymes), which, naturally, affects potential energy in that area. Many other dynamic processes also occur there, each of which leads to formation and transformation of potential wells. This processes occur simultaneously in many places and in a very large number of objects. Each of these objects can potentially become—at some point of time—a catalyst in effect for nuclear synthesis and transmutation (i.e. actually become a single-use micro reactor).

In the process of culture development, peculiar dimensions and shapes of these potential wells will inevitably (over some period of time) become close to optimal conditions, which creates a basis for synthesis and transmutation of elements.

This process of continuous "structuring" with inevitable "hitting"—by characteristic dimensions of micro distortions *R*—target optimal values R_0 at different points of time, inevitably engulfs all without exception areas of a growing microbiological culture. This circumstance makes a principal difference between a growing micro culture from almost all static crystals of palladium and titanium, where dimensions, shape and quantity of micro distortions are practically constant and there is no "adjusting" mechanism to optimal conditions.

If a nutrient medium, where a microbiological culture grows, has all the initial isotopic components necessary for transmutation, their falling into optimally-structured micro cavities causes initiation of synthesis reactions with generation of an isotope, which did not exist in the nutrient medium initially, and which is required for a culture's growth. This isotope is immediately assimilated by a growing culture and becomes part of its structure. That process is replicated over the entire area of growth. If a randomly generated isotope is not required for a culture's growth, it can, apparently, fluctuationally disintegrate into basic components.

CHAPTER 5

Essential Biochemical Factors of the Process of Transmutation of Isotopes in Biological Systems

In the previous chapter we discussed some of the possible molecular physical mechanisms which can explain and substantiate the running of low temperature nuclear reactions in optimized quantum mechanical systems. Essentially it was an attempt to give an answer to the question "How a low energy synthesis and transmutation reaction can be accelerated and optimized?"

An answer is based on the laws of quantum mechanics and—to an equal degree—can be attributed to any animate and still objects.

While analyzing the problem of transmutation of isotopes in growing biological cultures (especially the case of transmutation with generation of isotopes of such chemical elements which are not required by a growing culture in normal conditions) many additional specific questions arise.

The most important of them: "Why a growing culture needs this?; How is it accomplished?; Can this process be controlled?"

Distant future will provide full and final answers to these questions. The authors recognize all complexity of this problem. Our thesis below can be interpreted only as one possible solution based on our understanding of the problem.

Keeping consistent with our position of objective regularity of such a process we must note that an explanation ought to be sought among the known laws of physics, chemistry and biology.

5.1 Elementary constitution of biological objects and the problem of controlled synthesis in a growing culture

Let's first consider general patterns of the metabolism process in live systems related to their elementary makeup.

It is well known that any live system is a self-stabilizing object whose chemical elementary constitution is determined by its genetic code and stays practically unchangeable throughout the whole lifetime. Using the relative concentration classification as the basis chemical elements are usually divided into vital elements and necessary microelements.

The following elements are regarded as vital: O (typical concentration in a living culture about 24%), H (about 64%), C (about 9%), N (about 0.13%).

The following microelements are required for growth of biological cultures: Na ($7 \bullet 10^{-3}$%), K ($4.5 \bullet 10^{-2}$%), Ca ($7.5 \bullet 10^{-2}$%), Mg ($2 \bullet 10^{-2}$%), Fe ($8 \bullet 10^{-4}$%), P ($1.3 \bullet 10^{-2}$%), Si ($3.5 \bullet 10^{-2}$%).

The peculiar nature of life and growth of biological objects also requires the presence of some trace microelements Cl ($7 \bullet 10^{-3}$%), Al ($6 \bullet 10^{-3}$%), B ($6 \bullet 10^{-4}$%), Zn ($3 \bullet 10^{-5}$%), Li (10^{-4}%), Cu (10^{-5}%), Sr (10^{-5}%), Ba ($5 \bullet 10^{-6}$%), F ($3 \bullet 10^{-5}$%), Br ($6 \bullet 10^{-6}$%), Rb ($4 \bullet 10^{-6}$%), Sn (10^{-6}%), Ni ($5 \bullet 10^{-6}$%), Mo (10^{-6}%), Co (10^{-6}%).

For some microbiological cultures, microelements of S, Mn, J, Hg and others are also necessary.

Quoted values are average and may vary by several orders in some cultures.

How well balanced and mutually stable a correlation of chemical elements in an organism should be? What would happen if one of the elements is missing?

These are not idle questions. The very phenomenon of low energy transmutation of chemical elements and isotopes in biological systems and creating conditions for sustaining it is lodged upon the heuristic proposition that if some of the required elements or microelements is not present in the living environment (or nutrient media) than given that certain pre-requisites are met it will be synthesized as a result of the transmutation. In fact such an approach unambiguously suggests that the ratio of all the necessary elements in each type of living organisms is fixed.

A consistent study of this axial question was initiated in the founding works of Vernadski (Vernadski, 1931, 1987, in Russian). Without stumbling over the problem of consistent balance of elements in all live objects we will investigate it on the example of most simple objects—microorganisms. A convincing proof of consistent balance of elements was provided in the works of Syroezhkin (Syroezhkin, 2002, in Russian) dedicated to the study of mutually dependent metabolism of metals in microbiological cultures and more complex plants.

Fig. 5.1 gives a relative content of different metals in desiccated cells of microbiological cultures *Bacillus subtillus*, *Mucobacterium tuberculosis* and *Rhodococcus rodochrous* and in a more complex plant of the lily type *Allium aflatuense* as well as deviations from average values for the series of simultaneously conducted reference experiments.

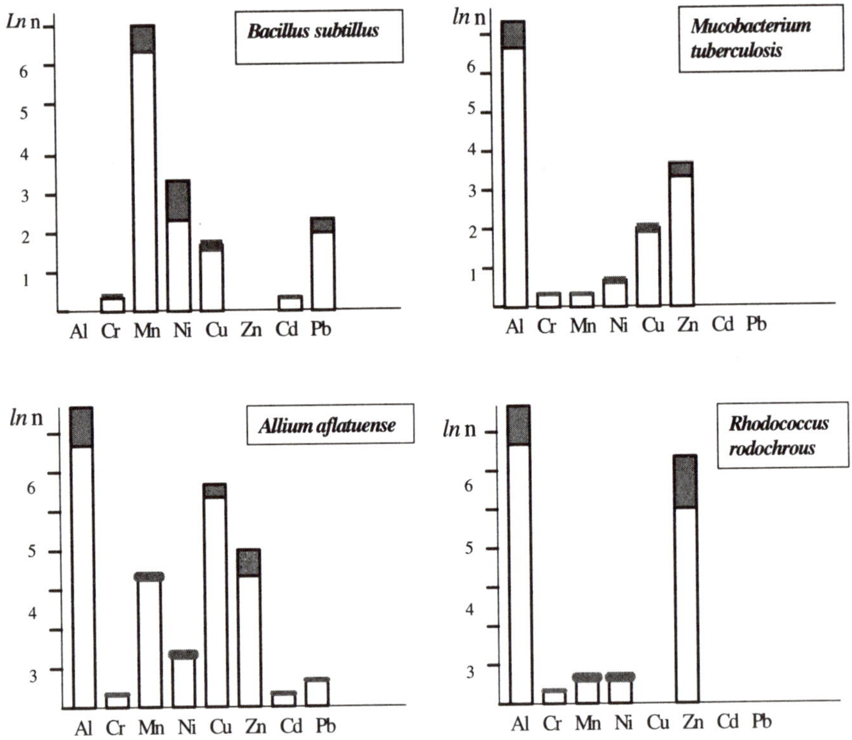

Fig. 5.1 Mass spectrum of desiccated cells of microbiological cultures *Bacillus subtillus*, *Mucobacterium tuberculosis* and *Rhodococcus rodochrous* and a more complex plant *Allium aflatuense*

It can be incurred from the data that the ratio of various metal microelements is constant for a given biological object but it varies greatly in different objects. Syroezhkin reported in his work that a deliberate and very significant increase of one of the metal's concentration in the nutrient medium did not cause the ratio of concentrations of various metals in the nutrient medium to change. For example, increasing the concentration of manganese by 10 times (from 1mg /ml to 10 mg/ml) did not change the overall specter. The specter of microelements itself is quite interesting (in particular, the presence of a relatively large quantity of aluminum or lead in cultures).

Moreover the ratio of various microelements changes dramatically at transformation from one stage of the life cycle of the same culture to another (changing physiological states).

On Fig. 5.2 there is the spectrum of chemical elements of the spores of the same culture *Bacillus subtillus* which are generated in cells during the full processing of glycerin contained in the nutrient medium. In order to stimulate such spore generation the exposure time for cells of Bacillus subtillus was increased from 15 hours (which corresponds to the data in Fig. 5.1) to 14 days before all glycerin was completely extracted.

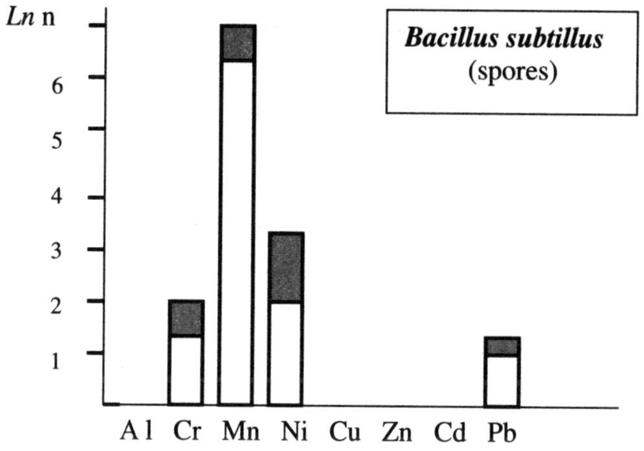

Fig. 5.2. Mass spectrum of desiccated spores of microbiological culture *Bacillus subtillus*

Comparing Fig. 5.1 and Fig. 5.2 we can conclude that changing the physiological condition of culture *Bacillus subtillus* has caused a significant alteration of the micro elementary content (specifically there is an almost complete elimination of cuprum).

According to the authors opinion (Syroezhkin, 2002, in Russian) it is due to changing the content of membrane enzymes (Basolo, 1971, in Russian; Hues, 1983) and transporting inter-faze membrane systems of a cell which occurs during the change of the physiological conditions of a culture.

These results reveal a non-trivial nature of interactions of different microelements. Changing the makeup of the nutrient medium it is possible to control the speed of a culture's growth. Lacking at least one of the microelements in the nutrient medium hinders the development of the entire biological object.

Apparently this mechanism of sustaining consistent elementary makeup of microorganisms is the key to answering a question posed earlier "why a growing culture needs the process of synthesis and transmutation?"

In this light a possible mechanism of culture growth stimulation becomes more evident as well as perhaps a mechanism of synthesis and transmutation stimulation. In particular the effect of synthesis stimulation of isotope Fe^{54} produced by adding cesium to the nutrient medium discovered by us may have a simple explanation.

Cesium as well as iron (or one of the biochemical analogues of cesium) are among the microelements required for sustaining growth provided that their optimal concentrations are regulated by quite rigid conditions. The absence of both components (or their low concentration) slows culture's growth.

Let's assume that in a growing culture there exist all the necessary conditions for synthesis of a particular iron isotope nuclei based on the effect of transmutation of phosphorus and natrium. Synthesis of iron only (with no cesium present) does not lead to getting synthesized nuclei involved in the metabolism process. The process of nuclei synthesis on the basis of the effect of dynamic suppression of the Coulomb barrier in an optimal potential well examined above is principally reversible in time before relaxation processes take effect. This circumstance leaves the door open to a reverse disintegration of an iron nuclei being synthesized into initial components (phosphorus and natrium).

Adding cesium causes iron Fe^{54} being synthesized to become immediately involved into metabolism process where it participates in a rapid generation of new electronic inter-atomic links in a potential well where synthesis takes place which accordingly alters the structure of the well. Since characteristic "electronic time" required for changing the structure of a potential well and establishing such links is . $\tau \approx \hbar^3/m_e c^4 \approx 10^{-17}$ sec—several orders shorter than "ionic time" (equal, in terms of the value's order, to the period $T \approx 10^{-13}$ sec of debaev oscillations of heavy ions in a condensed environment)—a reverse disintegration becomes impossible. In the end synthesized iron is integrated with the help of electronic links into a growing culture where it thus becomes fixed.

5.2 Specific features of microorganisms' interaction with metals

All biochemical processes are one way or another related to interaction of ions. Obviously the process of transmutation must also be based on such interaction. In this chapter we discuss specific features of microbiological cultures' interaction with various metals. The process of generation of ions occurs most naturally in metals. We shall hereafter include oxidized forms of metals in the term "metals" (with valency below zero) in the form of cations, anions, complex compounds, oxidants of hydroxides. This section is based on the Tashyrev's review (Tashyrev, 1995, in Russian).

First we shall provide a general classification of interaction processes with the intent of identifying and defining processes potentially capable of supporting nuclear transmutation.

In these works (Iske, 1989; Silver, 1984) it is shown that the process of interaction between ions of metal and microorganisms can manifest itself in following ways:

(1) mobilization (transferring of insoluble forms of metals into soluble);
(2) immobilizing (formation of insoluble sediments from soluble metal compounds);
(3) accumulation (sorption of metals on the surface of a cell or accumulation of metal inside a cell);
(4) formation of airborne metal compounds.

Very different mechanisms may provide foundation for each of the indicated types of interaction. Particularly in interaction of microorganisms with Hg^{2+} airborne compounds may be formed as mercury fumes Hg0 (Tegika, 1976) or methyl derivatives Hg^{2+}—monomethylmercury and dimethylmercury (Pan-Hou Hidemitsu, 1982). Both metal mercury and methyl mercury evaporate in water solutions but in the former case there is recovering transformation of metal (Hg^{2+} + 2e = Hg0) and attaching of methyl radicals to cation Hg^{2+} without changing the degree of its ability to recover

$$Hg^{2+} + CH_3 = Hg_2{+}CH_3;$$

$$Hg^{2+}CH^{3+} CH3 = (CH_3)_2Hg^{2+}$$

in the latter case. Immobilization of metals can be achieved by microorganisms through interaction of metabolites with metals or their own recovering transformation. Thus sulphate recovering organisms produce hydrogen sulphide which in reaction with many metals forms insoluble sediments. In interaction of microorganisms with chromates easily soluble at pH values close to 7, their recovering transformation takes place with generation of chrome hydroxide.

On the other hand the same transformation mechanism can produce totally different results. For example recovering transformation of mercury leads to generation of airborne metal mercury while during recovery of tellurides inside cells of bacteria metal tellurium crystals are formed. In this case the work of the same mechanism—the recovering transformation of metal—has led to formation of an airborne metal compound and accumulation of metal.

Let's investigate the first three types of microorganisms' interaction with metals more closely (the interaction process related to formation of airborne compounds occurs only at the presence of mercury ions).

5.2.1 *Metals mobilization*

Mobilization of insoluble compounds of oxidized metals occurs mainly because of decreasing pH of the environment during oxidizing of organic compounds by microorganisms as well as generating certain compounds by them which form organic-mineral compounds with metals called chelates. Most metals assume their soluble forms at reducing the pH of the environment to 3...4. One of the chelates' features is their ability to exist in an environment with a very wide pH range. If a cation Ti^{3+} begins to form sediment (titanium hydroxide) at pH equal to 2–2.5 adding chelate (citrate) to the solution makes titanium stable in the solution in the form of titanium citrate at a heavily alkaline reaction of the solution (pH equals 9...10).

Almost all microbial exometabolites can serve as chelating agents of metals—vitamins, enzymes, antibiotics, humic acids (Rohde, 1963), humus-like polymers—melanins (Senesi, 1987) and organic acids (Callander, 1983).

Mobilization of metals is most effective in anaerobic conditions during fermentation of carbons and natural polymers (polysaccharides) because there is a combination of two factors contributing to rapid dissolution of metals.

First during oxidizing of a substratum the environment pH decreases to 3...4 and second the main final products of fermentation (organic acids) are powerful chelates. In general chelation can increase solubility of oxidized metal compounds in natural conditions by 4 orders (Callander, 1983).

After transferring metal into a soluble form it becomes accessible to cells and becomes involved in microbial metabolism processes which can eventually lead to immobilizing, accumulation of metal or formation of airborne metal compounds. There is no information about existence of such metals in the literature which oxidized forms cannot be transferred into a soluble form at their interaction with microorganisms. It is evident that although mobilization itself does not causes nuclear transformations it can be an indirect catalyst for synthesis of elements and isotopes by stimulating microbial metabolism.

5.2.2 *Immobilizing of metals*

Precipitation of dissolved metals caused by microorganisms is observed most frequently during interaction with biogenic hydrogen sulphide generated during dissimilationary sulphate reduction or decomposition of organic compounds containing sulfur in its maximally restored state (for example cysteine, methionine). With an exception of single-charge alkaline metals of the main subgroup of the first group of elements (cationes of lithium, natrium, potassium and others) and several cationes of the second group of elements (calcium, magnesium), all other metals form predominantly insoluble sulphides of metals. As a rule no direct contact with sulphide gene cells is necessary for precipitation of metals (Ilialetdinov, 1982, in Russian) but this process is often combined with transporting cationes of metals to the cytoplasm of a cell and subsequent formation of sulphides inside a cell (Jones, 1976). Thus several species of *Desulfovibrio* and *Desulfotomaculum* actively accumulate iron in cells in the form of sulphide.

Metals can also be immobilized by interacting with proteins producing strong insoluble compounds (Weege, 1989). Thus kind of interaction is characteristic of copper. Carbon dioxide (CO_2) produced by virtually all heterotrophic microorganisms leads to formation of insoluble compounds of cadmium and lead ($CdCO_3$, $PbCO_3$) (Babich, 1979). Lead and other metals can form insoluble compounds with amber acid (Aickin, 1979). Beside that immobilization of metals is possible through interaction with peroxides (Wood, 1983).

5.2.3 *Accumulation of metals*

Microorganisms accumulate metals by depositing them on the surface of a cell or inside it. Concentration of metals in the biomass of microorganisms can exceed concentration in the ènvironment by several orders. For example coefficients of metals extraction by microorganisms in natural conditions (sea water) are the following: cadmium—910, cobalt—4600, copper—7000, iron—87000, lead—41000, manganese—94000, titanium—20000 and zinc—65000 (Bowen, 1966). Coefficient of absorption of nickel by some types of anaerobic microorganisms is equal 3000 (Lundie, 1988).

Accumulation of metals by microorganisms occurs not only at low concentrations of metals in water solutions but also at very high concentrations. Thus the culture *Thiobacillus ferrooxidans* ATCC 19859 is capable of growing and accumulating metals at concentration of cations of copper and zinc equal to 30 and 40 g/l respectively (Brahmaprakash, 1988). Concentration of metal bound with microorganisms' cells is most frequently lies within the range of several tens of milligrams to 1 gram of dry cellular mass but sometimes this value can amount to hundreds of milligrams to 1 gram. *Escherichia coli* cells accumulate zinc in concentration equal to 16 mg to 1 gram of dry cellular mass, and *Pseudomonas cepaciae* can absorb cadmium and lead in concentrations of 77 and 53 mg/g respectively (Drapeau, 1983). Microorganisms stable to influence of metals demonstrate an increased capacity to cell accumulation: *Alcaligenes eutrophus* CH 34 absorb from 130 to 150 mg/g of cadmium (Diels, 1987) while the culture *Pseudomonas sp.*—up to 200 mg/g of silver (Glombitza, 1988).

The same microorganism culture can efficiently accumulate metals related to different groups of elements. For example *Rhodotorula glutinis* accumulates manganese, iron, copper, zinc, cadmium, lead, beryllium and chrome in high concentrations (Drapeau, 1983). On the other hand the same metal can be accumulated by representatives of various taxons of microorganisms. Thus cadmium is absorbed by green micro alga (Sacaguchi, 1979), pseudomonads (Kurek, 1989), fungi (Gadd, 1983) and others.

Such a pronounced non-specific nature of accumulation of metals by microorganisms points at the predominance of physical and chemical mechanisms of microorganisms' interaction with metals. Indeed microorganisms most actively absorb metals during the first five minutes after adding metals to the suspension of microorganisms. Accumulation of metals occurs in most cases through their deposition on cell surfaces. Metals can be tied by capsular polysaccharide (Scott, 1988) provided that several metals can simultaneously accumulate in the same capsule. For instance cells of *Bacillus megaterium* ATCC 19213 forming a large capsule accumulate cations of copper, mercury, silver, iron, zinc and manganese (Cassity, 1984).

Accumulation of metals on the surface of cells can take place as a result of formation of a complex of metals with cell membrane proteins.

For example *Mycoplasma campricolum* accumulates iron which localizes in a cell membrane in the form of a complex with protein (Bauminger, 1980). Many metals form insoluble compounds on the cell surface such as $PbHPO_4$ (Aickin, 1979), $CdHPO_4$ (Macaskie, 1987), Au0 and $Pd(OH)_2$ (Beveridge, 1978), sulphides of metals and other compounds. Creation of insoluble compounds on the surface of cells is sometimes related to enzymes' activity. Thus formation of cadmium phosphate is associated with phosphatase's activity being induced by glycerol-2-phosphate and catalyzing its decomposition with generation of $HPO_4{}^{2-}$. This anion interacts with Cd^{2+} with generation of insoluble $CdHPO_4$ linked to cells (Macaskie, 1987).

Accumulation of metals on surfaces of microorganisms is also achieved through tying up by cellular walls' structures (Beveridge, 1978, 1982).

Metals can be tied up not only by cellular walls' structures but also accumulate inside cells. There is data related to accumulation of silver inside cells (Glombitza, 1988), zinc (Kasan, 1988), cobalt (Belov, 1985, in Russian), cadmium (Gadd, 1983), as well as some other metals. Evidently accumulation of metals inside cells is related to the work of transporting systems given that both active and passive transporting systems can take part in metal absorption (Ladoga, 1985). After penetrating inside a cell metals can become tied to proteins of the cytoplasm and inner membrane structures or form insoluble products inside a cell (Laube, 1980).

Generation of inner cellular insoluble metal compounds can be accomplished by very different mechanisms. For example oxidized telluride compounds are transformed inside a cell with generation of elementary telluride along the recovery route:

$TeO_3{}^{2-} + 3H_2O + 4e^- = Te^0 + 6OH$ (Gordgyan, 1970); cadmium and iron inside sulphate reduction cells form corresponding sulphides (Jones, 1989), while manganese can generate insoluble crystalloid $MnHPO_4$ in cytoplasm of cells (Kihn, 1988).

Despite the fact that in most cases accumulation of metals by microorganisms has a clear "topography" (localization), sometimes metals are simultaneously distributed in cell's walls, membrane and cytoplasm fractions (Kihn, 1988).

To our opinion the process of accumulation is directly related to the phenomenon of transmutation. This circumstance has been discussed earlier and was explained by the fact that several necessary conditions must be met in order to produce nuclear synthesis and transmutation, one of which was the need to get a synthesized isotope involved in the metabolism process as quick as possible, accompanied by its fixation as a stable nucleus (stable atom). It is obvious that biochemical processes contributing to the process of accumulation also promote nuclear transmutation.

5.3 Biological significance of microorganisms' interaction with metals

Biological significance of microorganisms' interaction with metals is determined by metal concentration, degree of its toxicity as well as metabolic potential of microorganisms.

With respect to any element three parameters of its concentration can be determined at which:

(1) metal is required for sustaining normal cell metabolism and is actively absorbed from the environment with the help of transporting systems;
(2) metal is contained in excess quantity and inhibits metabolism while cells activate their protective mechanisms which compensate the negative influence of metal;
(3) a lethal dose of metal irreversibly suppresses cell development.

We shall demonstrate it on the example of copper. Copper is part of the composition of blue proteins of methylotrophic bacteria containing copper—amicianin and azurin—which play the role of initial acceptors of electrons from methyl-amino-di-hydro-genase and transport electrons to cytochrome CH and CL (Lawton, 1985; Tobari, 1983). For most microorganisms stimulating concentration of copper lays within the narrow range of 0.1–0.2 mM when it is necessary for methylotrophic cells for an efficient functioning of copper-containing transporters.

At higher copper concentrations up to 0.5–0.8 mM its inhibiting properties are manifested. Damaging effects of copper are numerous. It is expressed in blocking metabolic processes in a cell (Towincr, 1976) (for example inhibiting transporting electrons in the respiratory chain (Smith, 1980), inhibiting glycolysis and tricarboxylic acids (Domek, Mattew, 1987); altering the transmembrane potential of cytoplasmatic membrane and subsequent puncturing of the membrane by its own electric field (Lebedev, 1989); formation of peroxide radicals during copper's reaction with thiol-containing components of a cell (Naiki, 1980)).

In response to the damaging effect of copper a cell can synthesize capsular polysaccharides and low-molecular proteins tying excess copper cations; superoxide-di-mutase, decomposing peroxide radicals (Naiki, 1980); lower the rate of copper absorption from the environment due to a mended structure of the cytoplasmatic membrane (Gadd, 1984). Further increase of copper concentration to 1–2 mM causes a 60% decrease of live cells within several minutes, and then—to a culture's destruction (Jones, 1989).

Therefore copper which is considered a toxic metal, can have stimulating moderately inhibiting and bactericidal effect on micro-organisms depending on its concentration.

Any type of transformation of metals inside a cell is possible only at the presence of efficient transporting systems and at a first glance it is doubtful whether we can expect such systems to exist in relation to metals-"xenobiotics". However most of them are stereochemical analogues of ions, necessary for normal life-sustaining processes of a cell (Ca^{2+}, Na^{+}, K^{+}, etc.) actively interacting with transporting systems. Thus oxygen anions of the VI group of elements (molibdate, chromate, selenate and tungstate) due to their stereochemical similarity with sulphate compete with the latter for enzymes and a transporting system (Tegika, 1976), while Co^{2+} and Cu^{2+} contend for transporting systems for zinc to microorganisms' cells (Kasparova, 1989, in Russian; Macaskie, 1987).

With a higher degree of adaptation of microorganisms to elements which were not found in significant quantities in the biosphere before we can observe their increasingly complete involvement in biological processes up to appearance of a closed cycle similar to circulation of carbon sulfur and nitrogen.

Indeed we can hardly consider selenium a xenobiotic or a toxic element with regard to microorganisms if they naturally perform all types of transformation of selenium (Doran, 1982; Oremland, 1986):

(1) oxidizing and reconstruction of non-organic compounds;
(2) immobilization in microbial cells and mineralization of organic compounds;
(3) methylation and demethylation of organic and non-organic compounds.

Possibly a similar adaptation of microorganisms occurs for toxic metals as well. On the other hand microorganisms lacking specific protective systems against metals and other toxic elements are capable of interaction with them with generation of elementary forms due to effects of reconstructive transformation. For example homogenate of cells *Micrococcus lactilyticus* (*Veillonella alcalescens*) reconstructs oxidized arsenic, bismuth, selenium, tellurium, lead, thallium, vanadium, manganese, iron, copper, molybdenum, tungsten, osmium, ruthenium, gold, silver and uranium provided that many of them can be reconstituted to zero valence (Woolfolk, 1966).

Therefore metals including highly toxic ones can be involved in metabolic processes without any "tragic" consequences for micro-organisms. Microorganisms' interaction with metals is manifested through its final result as mobilization (transformation of metals into a dissolved state) or to the opposite as immobilization—accumulation (transformation of metals into a insoluble state or their tying by cells).

Apparently the radio nuclides just as non-radioactive metals will form sulphides of metals during biogenic sulphate reduction. In other words radioactive metals (radio nuclides) should be regarded first of all as elements reacting with microorganisms in the same manner and through the same mechanisms as non-radioactive metals.

5.4 Biochemical analogues of vital microelements, metabolism of microorganisms and the question of controlled transmutation

While we could come up with a simple answer to the question: "Why a growing culture needs the process of synthesis and transmutation?"—it would be objectively required for replenishing chemical elements required for an etalon combination of chemicals—an answer to the next question: "Where and how does a culture do it?" requires investigation of actual mechanisms of interaction.

Back in the 30's, academician V.I. Vernadskiy suggested, that live objects are concentrates of chemical elements (Vernadskiy, 1987, in Russian). It is possible, that, primarily, stereochemical similarity of "macroelements" and "minor" elements, dispersed in the biosphere comprises the foundation of this phenomenon.

The proposed model of dynamic brief "removal" of the Coulomb barrier effect on the probability of nuclear synthesis is based directly on dimensional quantum processes, where quantum potential wells, optimized in size, play the central role. These dimensions are very specific for each type of interacting particles. Our physical model was an idealized version of the real physical or biochemical process.

In terms of ideas, the concept of interaction of microorganisms with metals developed by Tashyrev, also based on dimensional effects, is very close to the beliefs of the authors (Tashyrev, 1994, in Russian). It includes the following aspects:

1. Each metal has its peculiar oxidizing-reconstructive and(or) stereochemical analogue among vital microelements, adapt to the particular culture (or "macro elements" according to the terminology used) (Tashyrev, 1995, in Russian), which are well represented components in microorganisms' metabo lism (K^+, Na^+, Mg^{2+}, NO^{3-} and others). The oxidizing-reconstructive analogy, or redox analogy, imply proximity or parity of redox potentials of reconstructive reactions of a metal and a macro element; stereochemical—proximity or parity of ionic radiuses of a metal and a macro element.
2. One of the consequences of the stereochemical analogy and redox analogy is non-peculiar nature of interaction of microorganisms with metals. Receptor, transporting and oxidizing-reconstructive systems of a microbial cell make "errors" due to proximity of values of ionic radiuses or redox potentials of a metal and a macro element, and therefore a metal is either accumulated by a microbial cell or it is reconstructed (often with generation of insoluble in water compounds). Some metals are capable of both accumulation and reconstruction by microorganisms.
3. Due to the non-peculiar nature of the aforementioned processes, metals should interact with microorganisms, not adapted to them.

Stereochemical analogy presumes proximity or parity of ionic radiuses of a metal and a macro element. Microorganisms react with metals because of this stereochemical analogy. Receptor and transporting systems of a microbial cell make "errors" due to parity or proximity of values of ionic radiuses of a metal and a macro element and, hence, metal is accumulated by a microbial cell in a non-peculiar way (by "non-peculiar" we mean accumulation of metals by microorganisms not adopted to them).

In Tashyrev's works (Tashyrev, 1994, 1995, in Russian), a general survey of published data on accumulation of metals by microorganisms is made, along with definitions of theoretical patterns of metals' accumulation in a microbial cell, due to their stereochemical analogy with macro elements.

Table 5.1 Stereochemical analogy of metals.

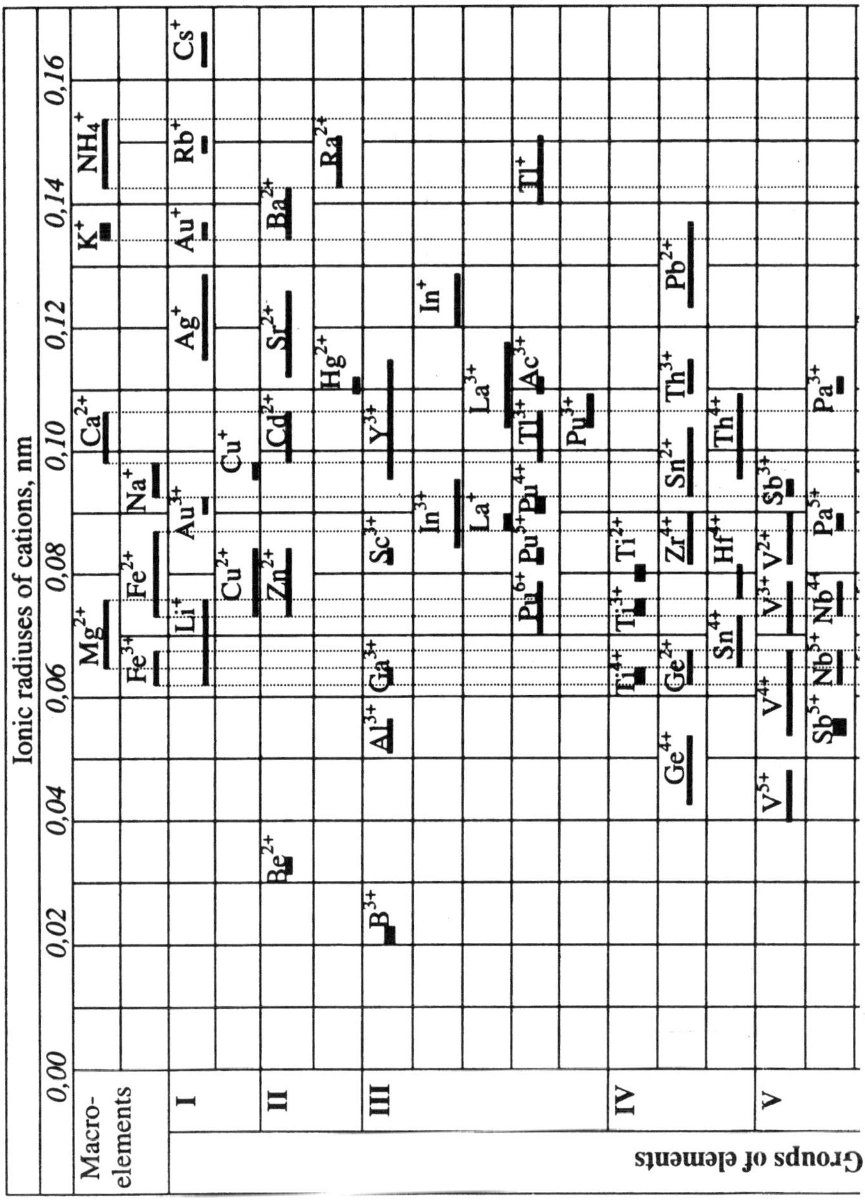

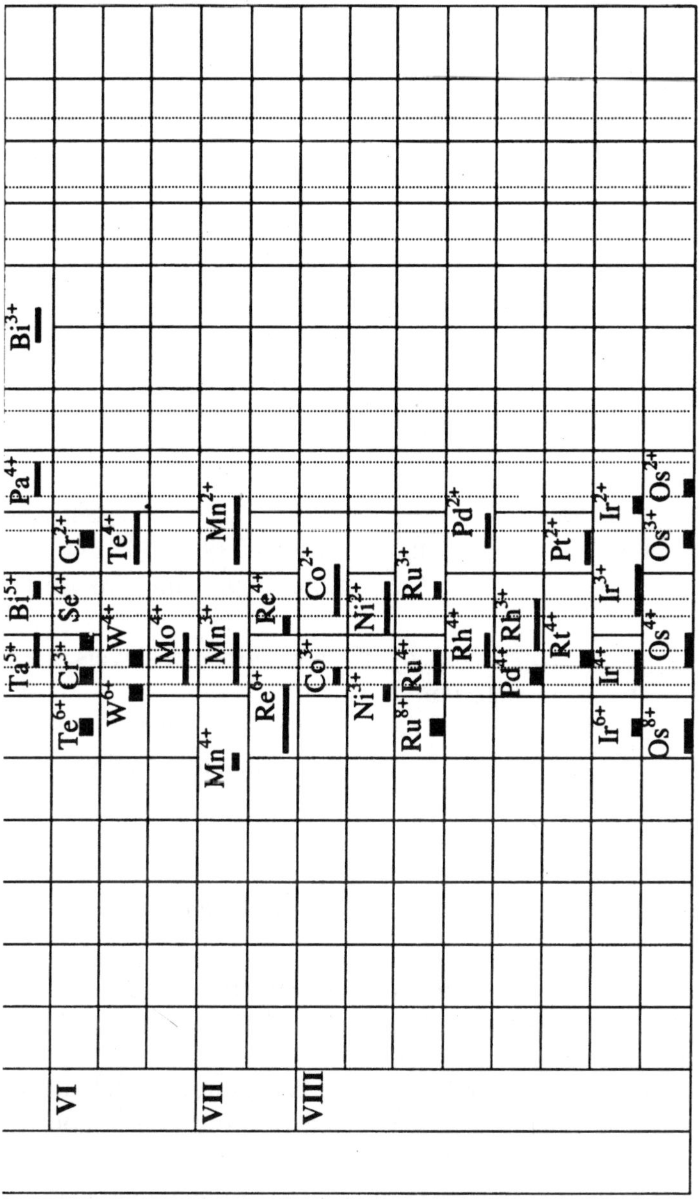

VI
VII
VIII
Bi3+
Pa4+
Ta5+
Bi5+
Te6+
Cr3+
Se4+
Cr2+
W6+
W4+
Te4+
Mo4+
Mn4+
Mn3+
Mn2+
Re6+
Re4+
Co3+
Co2+
Ni3+
Ni2+
Ru8+
Ru4+
Ru3+
Rh4+
Pd2+
Pd4+
Rh3+
Rt4+
Pt2+
Ir6+
Ir4+
Ir3+
Ir2+
Os8+
Os4+
Os3+
Os2+

Table 5.1 gives values (in nanometers) of dimensions of cations' ionic radiuses for macro elements and metals (Lurieaux, 1979).

The two upper rows give the dimensions of principal "macro elements"—Mg^{2+}, Ca^{2+}, K^{+}, NH_4^{+}, Fe^{3+} and Na^{+}. Other metals are situated according to their belonging to groups of elements of the periodical system of elements.

The size of the band, representing cation size, depends on differences in data, provided by various authors. For example, the ionic radius Mg^{2+}, according to Goldschmidt, Poling, as well as Belov and Bokyu, is equal 0.078; 0.065 and 0.074 nm respectively. Therefore, the size of the magnesium ion is represented graphically by the band from 0.065 to 0.078nm (see table 5.1, left, in the upper row). For purposes of convenience, ion sizes of macro elements are projected downward by dotted lines.

Among macro elements, three groups of ions can be clearly distinguished, which have similar ionic radiuses, forming three zones of stereochemical analogy.

The first group of ions—Fe^{3+}, Mg^{2+} and Fe^{2+}; they comprise **zone one** of the stereochemical analogy with dimensions 0.06–0.083 nm. The second group consists of Na+ and Ca^{2+} (**zone two**) with dimensions 0.095–0.106 nm.

The third group is represented by K^{+} and NH_4^{+}; it covers **zone three** with dimensions 0.133–0.159 nm.

At a closer look at table 5.1, it becomes apparent, that practically all metals (with the exception of Be^{2+}) are stereochemical analogues of macro elements.

Most cations are located in zone one of the stereochemical analogy (Fe^{3+}, Mg^{2+} and Fe^{3+}). Thus, dimensions of these macro elements precisely match 47 cations of metals. Another 17 cations of metals are very close to this zone.

Therefore, dimensions of macro elements of the zone one of the stereochemical analogy are close to 64 cations of metals, belonging to seven groups of the periodical system of elements.

The second zone, represented by Na^{+} and Ca^{2+}, directly includes 11 catons, with another 9 close to it. Zone three (K^{+} and NH^{4+}) holds 5 cations of metals and 7 other cations are close to it. Some cations hold intermediate positions between indicated zones. For example, Sr^{2+} is located between the second and the third zones, and Cs^{+} is located right of the zone three. It is known however, that Ca^{2+} and Sr^{2+} (Borst-Pauwels, 1986; Wilkens, 1989) as well as K^{+} and Cs^{+} (Bossemeyer, 1989) have corresponding common transporting systems.

Therefore, stereochemical analogy is manifested through differences in ionic sizes within the limits of 1–2 ionic radiuses, although maximal competition for transporting systems of microorganisms takes place among ions, having different or close ionic radiuses (Norris, 1977).

It can be seen, that practically all metals have stereochemical analogues. While forecasting non-peculiar accumulation of metals by microorganisms, it is necessary to allow for physical and chemical conditions of existence of oxidized forms of metals in actual water solutions. A vast majority of metals (see table 5.1) are stable in the form of cations only in an acidic environment and they change their forms at neutral values of pH. For example, Ti^{3+} at values of pH close to neutral forms an insoluble hydroxide $Ti(OH)^3$, and Cr^{6+} is transformed into $CrO_4{}^{2-}$, Cu^{2+} at pH more than 3 is transformed into $Cu(OH)_2$ and so on. However, stereochemical analogy of cations, presented in table 5.1, must take place (manifest itself) if microorganisms react with metals in acidic solutions.

Some metals (Zn^{2+}, Ni^{2+}, Co^{2+}, Cs^{+}, Sr^{2+}) are stable in the cation form both in acidic and neutral, as well as slightly alkaline solutions.

In neutral solutions, metals such as Cr(VI), Mo(VI), W(VI) can change their form to anionic. However, in the anionic form too, they have their stereochemical analogues among macro elements (Table 5.2).

Table 5.2 Sizes of ionic radiuses of several anions

Anion	*Ionic radius, nm*	*Anion*	*Ionic radius, nm*
NO_3^-	0,189 – 0,257	SO_4^{2-}	0,295
SH^-	0,200	PO_4^{3-}	0,300
$HS0_4^-$	0,206	CrO_4^{2-}	0,300
CIO_4^-	0,236	MoO_4^{2-}	0,345

The ionic radius of CrO_4^2 for example, is equal to the orresponding ionic radius of PO_0^{3-} and very close to the size of the ionic radius of SC_4^{2-}.

Theoretical forecast of the non-peculiar microbial transport of metals due to their stereochemical analogy with macro elements matches, to a high degree, the experimental data.

Table 5.3 gives information about interaction of microorganisms with stereochemical analogues-metals.

Table 5.3 Microorganisms' interaction with macro elements and their stereochemical analogues

Stereochemical analogues		*Micro-organisms*	*Type of interaction in a biological system*	*Source*
Macroelement	*Analogue metal*			
1. K^+	Rb^+	*Blastocladiella emersonii*	Common transporting system of ions to a cell	Kropf, Harold, 1982
2. K^+	Rb^+, Ba^{2+}	*B. emersonii*	Common transporting system of ions from a cell	Van Brunt, 1982
3. K^+	Rb^+, Cs^+	*Escherichia coil*	Common transporting system of ions to a cell	Bossemeyer, 1989
4. K^+	Rb^+	*Anabaena variabilis*	Common transporting system of ions to a cell	Reed, 1981
5. K^+	Rb^+	*Saccharomyces cerevisiae*	Common transporting system of ions to a cell	Rodrigues-Navarro, 1984
6. Mg^{2+}	Co^{2+}, Ni^{2+}	*Klebsiella pneumoniae*	Common transporting system of ions to a cell	Ainthworth, 1980
7. Mg^{2+}	Mn^{2+}, Cd^{2+}	*Bacillus subtilis*	Common transporting system of ions to a cell	Burke, 1988
8. Mg^{2+}	Mn^{2+},Co^{2+}	*Escherichia coli*	Co^{2+} inhibits transporting Mn^{2+} to a cell	Laddaga, 1985
9. Fe^{2+}	Co^{2+}	*Acidophilium sp.*	Competition for transporting system of ions to a cell	Don, 1989
10. Ca^{2+}	Sr^{2+}	*Saccharomyces cerevisiae*	Common transporting system of ions to a cell	Borst-Pauwels, 1986
11. Ca^{2+}	Zn^{2+}, Cd^{2+}	*Escherichia coli*	Zn^{2+} activates transporting of Cd^{2+} to a cell	Sacaguchi, 1981
12. Ca^{2+}	Zn^{2+},Co^{2+}, Mn^{2+}	*Candida utilis*	Common transporting system of ions to a cell; Zn^{2+} inhibits transporting of Co^{2+} to a cell	Каспарова, 1989
13. Ca^{2+}	Mn^{2+},Cd^{2+}	*B. subtilis*	Common transporting system of ions to a cell	Laddaga, 1985
14. Ca^{2+}	Zn^{2+}, Cu^{2+}	*Debariomyces hansenii*	Competition for centers of tying metals in a cell's walls	Wakatsuki, 1970
15. Fe^{2+}	Co^{2+},Mn^{2+}	*B. brevis*	Fe^{2+} и Co^{2+} inhibit transporting Mn^{2+} to a cell	Jack, 1979
16. Fe^{2+}	Mn^{2+},Cd^{2+}	*B. subtilis*	Fe^{2+} и Co^{2+} inhibit transporting Mn^{2+} to a cell	Burke, 1988
17. Fe^{2+}	Mn^2H, Cd^{2+},	*Staphylococcus aureus*	Competition for transporting system of ions to a cell	Weiss, 1976

18. Na^+, Mg^{2+}, Ca^{2+}	Cd^{2+}	*B. cereus, E. coli, Aspergillus niger*	Increasing protective action of cations from Cd^{2+} in the following order: $Na^+>Mg^{2+}>Ca^{2+}$.	Naidu, 1988
19. Mg^{2+}, Ca^{2+}	Ni^{2+},Zn^{2+},Mn^{2+}, CO^{2+},Cd^{2+}	*S. cerevisiae*	Maximal competition for transporting system between ions with close ionic radiuses	Naidu, 1988
20. Mg^{2+}	Co^{2+},Mn^{2+}	*E. coli*	Common transporting system of ions to a cell	Park, 1976
21. Mg^{2+}, Ca^{2+}	Cu^{2+},Al^{3+}, Zn^{2+},Fe^{2+}, Cd^{2+}	*E. coli*	Suppression of absorption of copper by ions of metals and macro elements	Bairdy, 1980
22. Mg^{2+}	Ni^{2+}	Микроэлементы	Competition for common centers of cellular wall	Babich, 1981
23. SO_4^{2-}	CrO_4^{2-}	*Pseudomonas aeruginosa, P. fluorescens*	SO_4^{2-} activates transporting CrO_4^{2-} to a cell	Ohtake, 1981
24. SO_4^{2-}	WO_4^{2-}, MoO_4^{2-}	*Rhizobium sp.*	WO_4^{2-} activates transporting MoO_4^{2-} to a cell	Baxi, 1988
25. SO_4^{2-}	MoO_4^{2-}, CrO_4^{2-}	*S. cerevisiae*	MoO_4^{2-} and CrO_4^{2-} inhibits transporting SO_4^{2-} to a cell	McCready, 1979
26. SO_4^{2-}	MoO_4^{2-}	*Desulfovibrio desulfuricans, Desulfotomaculum orientis*	MoO_4^{2-} inhibits transporting SO_4^{2-} to a cell	Fukui, 1989
27. SO_4^{2-}	MoO_4^{2-}, CrO_4^{2-}, Wo_4^{2-}, SeO_4^{2-}	*Desulfovibrio sp.*	Anions of metals inhibit transporting SO_4^{2-} to a cell	Taylor, 1979
28. NO_3^-	MoO_4^{2-}, CrO_4^{2-}, WO_4^{2-}	*E. coli*	CrO_4^{2-}, NO_3^-, WO_4^{2-} inhibit reduction of MoO_4^{2-}	Bairdy, 1980

Apparently, regardless of taxonomic location of microorganisms, stereochemical analogues share the same transporting systems. In this manner, a common transporting system for K^+ and Rb^+ is found among mushrooms (*Blastocladiella emersonii*), yeast (*Saccharomyces cerevisiae*), bacteria (*Escherichia coli*) and cyanide bacteria (*Anabaena variabilis*). Molibdate, chromate and tungstate are the stereochemical analogues of sulphate, having common transporting systems in aerobic (*Pseudomonas aeruginosa, P. fluorescens*), optionally anaerobic (*Escherichia coli, Saccharomyces cerevisiae*) and obligatorily anaerobic

(*Desulfovibrio desulfuricans*, *Desulfotomaculum orientis*) microorganisms.

It is also worth noting, that the presence of a stereochemical analogue may lead to stimulation, as well as inhibition of transporting a metal to a cell. This is how, in the case of *P. aeruginosa* and *P. fluorescens*, sulphate activates transportation of chromate to a cell, and in the case of *Rhizobium sp.* tungstate activates transportation of molibdate. Another example. In the culture *S. cerevisiae* and sulphate-reducing bacteria, chromate and molibdate compete for the system of transporting sulphate to a cell.

From the data provided (see table 5.3), it can be seen, that, presently, interaction of microorganisms with metals included in zone one of the stereochemical analogy (Fe^{3+}, Mg^{2+}, Fe^{2+}) has been reasonably well studied.

Metals, that are stable in water in the cationic form and broad range of pH, also belong to this category: Zn^{2+}, Cd^{2+}, Mn^{2+}, Ni^{2+} and others. These metals have common transporting systems with macro elements (first of all, with Mg^{2+}) and, as a rule, compete with them. Competition for transporting and It would be rightful to say, that stereochemical analogy of metals and nonpeculiar accumulation of metals due to stereochemical analogy is a frequent phenomenon in the world of microorganisms. We must also pay attention to the fact, that the same stereochemical analogues (for example, Mg^{2+}, Co^{2+}, Ni^{2+}) share transporting systems with microorganisms, associated with different taxons (*Klebsiella*, *Escherichia*, *Bacillus*, *Alcaligenes*, *Aspergillus*; see table 5.3). Therefore, apparently, non-peculiar transporting and bioaccumulation of stereochemical analogues by microorganisms is a common pattern in general biology. Tashyrev was the first to notice that.

We should also mention, that, at the presence of organic acids and other exometabolites in the environment, metals form complex compounds with them, which dimensions are much greater, than dimensions of metal ions.

Nevertheless, the probability of cation size being the determining criteria of a stereochemical analogy is very large. Thus, if cation sizes are close S^{6+} (0.029–0.034 nm) and Cr^{6+} (0.035–0.053 nm), than the same elements are stereochemical analogues in their anionic state as well, because they have large ionic radiuses: SO_4^{2-} (0.295 nm), CrO_4^{2-} (0.300 nm). Apparently, the same effect will take place during generation of complex compounds with exometabolites of microorganisms.

Obviously, the accumulation of metals by microorganisms due to active transporting is effective in cases, when their concentration is comparable with "physiologically necessary concentration" of a macro element. Considering the fact, that the microorganism population's need for macro elements (and for microelements) is 10 mg/l...20 mg/l, we can assume, that non-peculiar microbial accumulation of metals will be realized within this range of concentrations.

At higher concentrations of metals (20–100 mg/l), their removal from water solutions is possible because of absorption by cellular wall (Beveridge, 1976; Beveridge, 1978; Brady, 1991; Marquis, 1976; Pooley, 1976) or by polysaccharide capsule (Ford, 1987; Gessey, 1988; Mittelman, 1985; Scott, 1987, 1988). If concentration of metals in a water solution exceeds 100 mg/l, neither active transporting within the cell, nor absorption by cellular structures will provide necessary degree of purification. In this case, effective removal of metals can occur only due to generation of insoluble metal compounds with exometabolites, for example, sulphides, phosphates or carbonates.

Systematization of metal ions from the point of view of stereochemical analogy, made by Tashirov, allows to hypothesize the presence of yet unknown common transporting systems for many metals. For example, by the size of ionic radius, Rb^+, Ba^{2+}, Ra^{2+} and Tl^+ are stereochemical analogues of Cs^+. Beside that, despite the fact, that Cs^+ is the analogue of K^+, it is closer to NH^+ by its size, that to potassium. However, competition between Cs^+ and NH_4^+ for transporting systems of microorganisms is still not properly researched. It is possible, theoretically, to discover transporting systems in microorganisms for the following pairs of cations:

Hg^{2+}—Sr^{2+}, Ca^{2+}—Sn^{2+}, K^+—Pb^{2+}, Na^+—Sn^{2+},

K^+—Ag^+, Ca^{2+}—Pb^{2+}, Ca^{2+}—Hg^{2+}, Na^+—Cu^{2+}

and others (see table 5.1). In several works, there is a demonstration of competition for transporting systems between anions of metals, belonging to the VI group of elements and sulphate (see table 5.3). However, the influence of phosphate (PO_4^{3-}), having the same ionic radius as sulphate, on transporting anions of heavy metals, is yet to be studied.

Stereochemical analogy of cations of metals is also manifested in enzymatic systems of microorganisms (see table 5.4). As a rule, macro elements and metals with close ionic radiuses activate enzymes and can substitute each other in active enzymatic centers.

Thus, macro elements of zone one of the stereochemical analogy—Mg^{2+} and Fe^{2+}—activate pyrophosphatase activity of microorganisms contained in soil. The same effect is produced by their structural analogues—Co^{2+}, Zn^{2+} and Mn^{2+}—which have ionic radiuses close to macro elements (#9, table 5.4). The ability of metals with close ionic radiuses to substitute each other in enzymatic systems without interfering with their functions is, apparently, a frequent phenomenon.

Table 5.4 Stereochemical analogy of metals in enzymatic systems of microorganisms

Stereochemical analogues		*Enzymatic system of a microorganism, influence of analogues*	*Source*
Macro-element	*Analogue metal*		
1. Mg^{2+}	Mn^{2+}	Activation of the adenylate cyclase system of *Escherichia coli*	Liberman, 1985
2. Ca^{2+}	Sr^{2+},Cd^{2+}, Mn^{2+}	Reconstruction of active glucosedehydrogenase Acinetobacte calcoaceticus from the apo-form	Wilkens, 1989
3. Ca^{2+}	Cg^{2+}, Mn^{2+}, Co^{2+}, Zn^{2+}	Activation of L-galactit-1-phosphatedehydrogenase *Klebsiella pneumoniae*	Markwell, 1981
4. Ca^{2+}	Mn^{2+}, Cu^{2+}	Substitution of copper by managanese in Cu/Zn-superoxidedismutase *Bacillus staerothermophillus*	Bannister, 1985
5. Ca^{2+}	Co^{2+}, Zn^{2+}	Re-activation of alcohol-dehydrogenase *Zymomonas mobilis* by ions of cobalt or zinc	Kinoshita, 1985
6. Fe^{2+}	Co^{2+}, Zn^{2+} Cd^{2+}, Ni^{2+}	Re-activation of creatin-imino-hydrolase *Flavobacterium filamentosum*	Esders, 1985
7. Mg^{2+}	Mn^{2+}, Co^{2+}, Zn^{2+}, Ni^{2+}	Activation of yeast hypoksantin/guaninphosphor-ibosil-trans-ferase by metal ions	Ali, 1986
8. Fe^{2+}	Mn^{2+}, Cu^{2+}	Substitution of copper in Cu/Zn-superoxide- dismutase of 8 strains o *Pseudomonas* and *Xanthomonas sp.* by manganese and iron	Steinman, 1985
9. Mg^{2+}, Fe^{2+}, Ca^{2+},	Co^{2+}, Mn^{2+}, Zn^{2+}	Activation of pyrophosphatases of 6 strains of soil microorganisms	Searle, 1977
Mg^{2+}	Zn^{2+}, Sr^{2+}, Co^{2+}	Re-activation of the apo-form of metalproteinase *Pseudomonas fluorescens*	Diermayr, 1987
10. K^{+}	Tl^{+}, Rb^{+}	Tying of Tl^{+} and Rb^{+} by yeast aldehydedehydrogenase, activated b potassium	Bostain, 1982
11. Ca^{2+}	Sr^{2+}, Cd^{2+} Mn^{2+}	Competition of ions for an active glucosedehydrogenase center	Wilkens, 1989

Thus, for example, Co^{2+} and Zn^{2+} can substitute each other in L-galactit-1-phosphate-de-hydro-genase, alcohol-de-hydro-genase, creatinimino-hydrolase, hypoxantin/guanin-phosphor-ibozil-trans-ferasa, pyrophosphatase and metal-proteinase in various microorganisms (#3, 5, 6, 7, 9 and 10 in table 5.4). For three enzymes, the activating metals are Mn^{2+}, Co^{2+} and Zn^{2+} (#3,7, 9 in table 5.4), etc. Therefore, stereochemical analogues of zone one are also physiological analogues.

The size of ions has, apparently, greater significanse for enzymatic systems, than ionic charges, although we could expect enzymatic systems to display greater affinity with cations with equal charges. Thus, carboxilase is activated both by double-charged Mg^{2+} and Fe^{2+} and by triple-charged Al^{3+} and Fe^{3+}. The same pattern is observed in the relationship between Mg^{2+}, M^{2+} and Al^{3+}, Cr^{3+} for phospho-gluco-mutase (Kalinin, 1971).

It is worth noting, that some enzymes can activate cations, which are quite different from each other by their sizes.

Thus, ions, which are not stereochemical analogues, activate malat-dehydrogenase: Mg^{2+} (zone one), K^+, Cs^+, Rb^+ (zone three) (Kalinin, 1971).

However, enzyme activation in these cases is determined, perhaps, not by cations of metal directly, but rather by other factors—ionic strength of the solution, the value of s-potential and other factors.

5.5 DNA reparation and the problem of transmutation of isotopes

One of the possible elements of a biological system where transmutation of isotopes can occur is a DNA macromolecule. This macromolecule holds genes, responsible for processes of self-recovery of a DNA itself and genes regulating the synthesis of protein.

It is known, that a DNA molecule consists of two complimentary chains, forming a coupled sequence of nucleotides (*adenin(A), timin(T), cytosine(C), guanin(G)*).

Under the influence of an external ionizing radiation or free radicals DNA chains can rupture. Most of all, it is related to macromolecules of DNA free of protein. Single ruptures are cured with the help of specialized ferment *ligase*. Paired ruptures can not be repaired by ferments and the process of DNA self-recovery can involve only purely physical mechanisms of self-reparation, considered in the following works (Vysotskii, Kornilova, Samoylenko, 1996, 1997; Vysotskii, Pinchuk, 1999; Vysotskii, Kornilova, Samoylenko, Pinchuk, 1999, 2001; Pinchuk, Vysotskii, Zhmudskii, 1999; Vysotskii, Pinchuk, 1999; Pinchuk, Vysotskii, 2001).

It was demonstrated in these works that the character of interaction between the end pairs of nucleotides, located on both sides of a double rupture intricately depends on the following several factors:

(1) width of a rupture L;

(2) the Coulomb interaction of charges q_{n_i} and q_{m_j}, distributed on the surface of the nucleotides i and j, taking into consideration the reflecting ability of ions of a water-salt inner cellular media, being in effect an electrolyte:

$$V_C(L) = \sum_{n_i, m_j} V_C(|\vec{r}_{n_i} - \vec{r}_j|)$$

(3) the dispersion features, characterized by dielectric permeability of the inner cellular liquid $\varepsilon(\omega)$ in the frequency range from quasi static $\omega_{\min} \approx 0$ to Roentgen $\omega_{\max} \approx c/L$ and dielectric permeability $\varepsilon_i(\omega)$ and $\varepsilon_j(\omega)$ of nucleotides.

Full interaction includes both the sum of Coulomb interactions distributed on the surface of charges of different pairs of nucleotides (formed by Del Re σ-electrons and Hukkel π-electrons providing for specific geometry of Watson-Krikov pairs) and the electrodynamic Van Der Waals dispersion interaction between these pairs of nucleotides in the inner cellular environment.

On the basis of performed analysis we found that for the four combinations of end pairs of nucleotides AT—TA, AT—GC, AT—CG, GC—CG (from six potential combinations AT—AT, AT—TA, AT—GC, AT—CG, GC—GC, GC—CG on the opposite sides of a double rupture) the character of interaction between the ruptured parts of DNA with any width of rupture L corresponds to their mutual attraction and the absence of a potential barrier. This result leads to a possibility of self-elimination of a paired rupture with restoration of former combinations between the end pairs of nucleotides generated by rupture and self-restoration of DNA's integrity. We also found, that for two special combinations of the end pairs GC—GC and AT—AT with the width of the rupture about $L_0 \approx 7$ A and pH ≤ 7–9 (as well as with relative concentration of hydrated electrons in the water-salt inner cellular liquid $\alpha \leq 10^{-7}$) a potential barrier $V(L, \eta, \alpha) > 0$ exists in the space between the ruptured ends of DNA, which prevents the process of self-reparation of the ruptures, which width exceeds L_0 (Fig. 5.3).

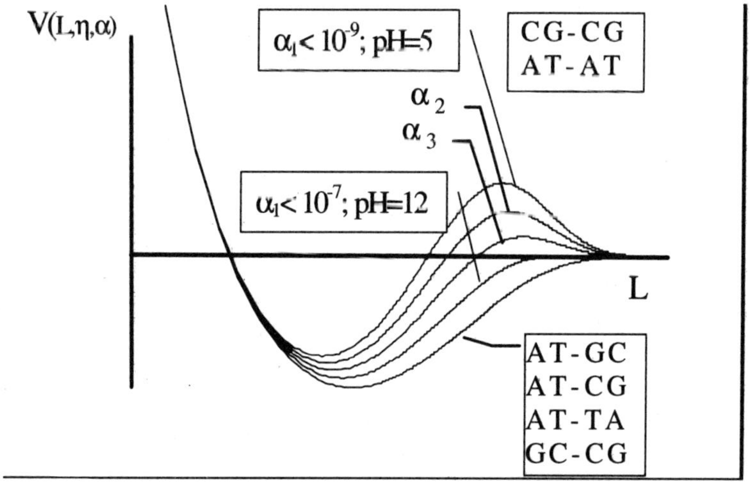

Fig. 5.3 Spatial structure of a potential barrier in the area of a double DNA rupture at various pH and different concentration of hydrated electrons in the inner cellular water-salt environment

The presence of this antireparation barrier and its height $V_{max} \equiv V(L_0, \eta, \alpha)$ depend on the relation between the Coulomb and dispersion forces, which, in their turn, are determined both by the type of nucleotides in end pairs and the characteristics of the inner cellular liquid (viscosity η., temperature *T*, ionic and salt content and concentration of hydrated electrons . and others). The value Vmax can significantly exceed the thermal energy *kT* (for example, for interaction of the end pairs of nucleotides GC—GC, we have $V_{max} \approx 3\ k_bT$.

At the presence of external radiation fields and other influences, due to changing characteristics of the inner cellular liquid, there is changing of characteristics of the interaction forces between the end pairs, deformation of the barrier $V(L, \eta, \alpha)$ and reduction of its height. With a certain combination of parameters for the environment and characteristics of external influence the barrier disappears for all possible types of paired ruptures, which leads to a possibility of a sharp increase of the probability of automatic reparation of all types of ruptures and a significant suppression of DNA degradation. In fact, such a process corresponds to a solution of the problem of mutual recognition and following self-reparation of separated fragments of a DNA biomolecule. It was demonstrated, that the process of forming and elimination of a rupture has diffusion nature and related to fluctuation influence of water molecules on the surface of separated parts of DNA. Direct calculations show that this process is sensitive to temperature of the inner cellular environment and characterized by a very long relaxation time (it can reach several hours).

The central part of a potential pit in the vicinity of a double rupture has a shape, which is close to parabolic, which can promote fulfillment of processes leading to a short-term suppression of the Coulomb repulsion of the nuclei of atoms, which may be located in the area of a rupture. Similar catalytic effect can occur in the area of action of a DNA-polimerase.

A closer analysis of the physical aspects of such processes was provided in the chapter dedicated to the theory of nuclear transmutations.

5.6 Adaptation of biological objects, analogy of microelements and synthesis of isotopes – the rule of biological evolution

Therefore, manifestation of the stereochemical analogy of metals and vital microelements, adapt to a culture, is the foundation of the effect of non-specific accumulation of metals in microorganisms.

Apparently, this process is the evolution's answer to the global dilemma – how to combine development and adaptation of biological objects, each one of which contains a genetically predetermined set of elements, with a random character and dissimilar distribution of elements in the outer environment, as well as constant environmental changes?

Perhaps, the very process of transmutation in growing biological systems with generation of necessary microelements or their analogues is just the sought answer of the evolution to this dilemma.

In our opinion, the process of transmutation occurs in places, where there is competition based on the stereochemical analogy (at least transporting and fermentation systems). The area, where this competition takes place determines the area, where transmutation itself is performed.

Can we point to a specific spot, where this ingenious nuclear reactor works? Possibly, there could be many such places (otherwise, reactions could be such rare events that they would be impossible to detect). Also note, that transmutation occurs with a higher probability in structural parts of biological objects, which are subjected to dynamic influences (zone of growth, non-stationary transport systems, dynamic response systems to any kind of agitation and so on).

It is also obvious, that the efficiency of the process of transmutation largely depends on whether the initial preconditions for a short-term elimination of the influence of the Coulomb barrier on nuclear interaction are fulfilled. It is necessary that the elements participating in the process of transmutation (isotopes) were in a soluble form (for metals, it corresponds to the process of mobilization). Clearly, for a higher probability of such processes, it is required that the initial isotopic components in soluble forms were separated from accompanying chemical elements (for example, nuclear interaction with participation of protons or deuterons, due to the presence of a varying Coulomb barrier, is executed differently, depending on whether these nuclei are part of water molecules or they are presented in the form of independent ions).

This requirement immediately leads to the need for detailed consideration of an entire array of oxidization-restoration processes and allowing for the appropriate redox potentials according to the scheme, which is analogous to the one considered earlier.

Finally, the outcome localization of products of transmutation depends on whether such product is adapted to a given culture. For adapted elements (for example, synthesized iron, which, apart from its isotopic content, does not differ in any way from natural iron) localization matches the natural and characteristic distribution for this particular micro element. Meanwhile, for alien, not adapted elements the final localization of the products of transmutation coincides with distribution of admixed, not adapted elements (sediments on he surface of cells, connected by capsulate polysaccharides, formation of a complex of metals with proteins of the cellular membrane, connection with structures of cellular walls, connection with proteins of the cytoplasm and internal membrane structures or formation of non-soluble products inside a cell and so on) analyzed earlier.

A precise search for a "nuclear bioreactor" can be conducted with the help of a position-sensitive precision analytical methods (for example, tunnel microscopy or high resolution mass-spectrometry).

CHAPTER 6

Prospects and Possible Methods of Using the Phenomenon of Transmutation of Isotopes in Biological Systems

6.1 Using the phenomenon of transmutation of isotopes for prevention and treatment of illnesses, etiologically caused by deficit of microelements

Before 19th century, there was a common belief, that all live organisms consist of a limited number (about 12–15) of chemical elements (so called macro elements). H, C, N, O, P, S, K, Ca, Fe and about 10 other "biogenic elements" were considered such elements. It was believed in the beginning, that macro elements are the lightest chemical elements. Then, this opinion was abandoned for another opinion, that macro elements are generated according to their ability to form compounds with carbon.

A principally new position on this issue appeared after the basic works of Vernadski on geochemical research of live matter. The first detailed analysis of a grandiose by its scale geochemical outcome of the Earth biosphere's activity, which—over the period of one billion years—has cardinally altered the entire biogenic structure of land and water on Earth, became the basis for this aspect of studying live nature. The biosphere has become the real generating and regulating factor of the nature's evolution. Throughout the process of evolution, the world has become an extremely intertwined and adapt to these processes. These changes were finally solidified in the genetic code. While the content of macro elements in various live organisms is roughly the same and the only significant differences are associated only with the structure of organisms, then the microelementary content varies very widely (without any visible reasons) from one kind to another, being rigidly fixed for each particular kind. These differences can be huge. While, normally, the content of microelements does not exceed hundreds and thousands of one percent, there are organisms, which accumulate and contain such heterogeneous chemical elements as Si, Fe, Mn, Al, Sr, Ba in quantities, comprising several percent! These are real enrichment plants with very high efficiency. In fact, each individual chemical makeup can be regarded as a specific feature of a given kind of live organisms (primary elementary quantization). On Fig. 6.1, there is distribution of the average (throughout the whole live matter of the Earth's land surface) concentration of various chemical elements (Vinogradov, 1935).

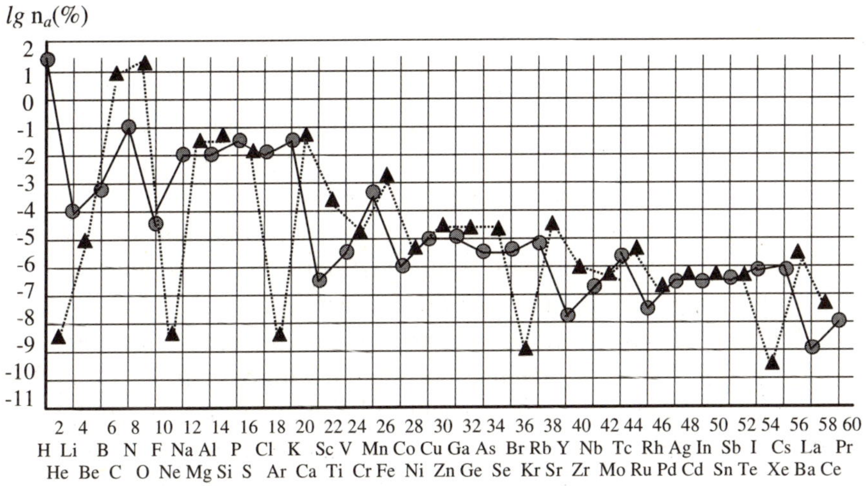

Fig. 6.1 Distribution of the average (throughout the whole live matter of the Earth's surface) concentration of various chemical elements

It can be seen from this Fig., that although the concentration of heavier elements in biological objects is actually decreasing (on average), distribution of elements with higher numbers does not decrease monotonously. It reflects the rules of evolution and functionality of certain microelements. Moreover, distribution of elements with even and odd numbers (even and odd number of protons in the nucleus) shows specific recurrence, ultimately associated with nuclear charge properties of nuclides.

This opinion was fully confirmed, when it was determined empirically, that the reason of many endemic diseases is the lack or excess of certain chemical elements (microelements) in the environment (soil, water, plants).

It was discovered, that practically all chemical elements should be contained in normally functioning live organisms. Both macro elements and microelements should be found in strict proportions (at certain concentrations) in an organism under normal conditions, providing its normal activity. Any deviations towards excess or deficit of the quantity of microelements (in relation to the optimal concentration of a given kind) cause pathologies, while significant deviations lead to illnesses.

A deficit or an excess of microelements leads to similar negative consequences, as in the case of vitamins. Apparently, it is not coincidental, since microelements are part of vitamins, enzymes and hormones and are capable of modifying their effects considerably. For example, thyroxin contains 65% of iodine, vitamin B12 contains 4% of cobalt, zinc is part of the enzyme carbonic anhydrase, the synthesis of haemoglobin is impossible without copper.

Each microelement is responsible for its own pathology. For example, the deficit of iodine leads to hypothyroidism with peculiar symptoms, while its excess quantity causes thyrotoxicosis; shortage of magnesium is manifested by general nervous malfunctions such as convulsions or rachitis, and its surplus is responsible for general depressed states of the human organism (inertia and sleepiness). There are very many examples of direct connection between various pathologies and violations of optimal microelementary makeup. We know only of those, which are correlated with violations of optimal makeup of the microelements, present in large quantities. If, on the other hand, the normal concentration of microelements is very small, such correlation is practically impossible to reveal. We can reasonably think, that much pathology, not presently attributed to violations, which are difficult to detect, are, in reality, caused by them.

There is one more aspect, which became apparent only after the discovery of the genetic code. While a balanced (on average) individual chemical composition may be regarded as a defining feature of a given kind of living organisms (primary elementary quantization), individual features of each organism within such kind, recorded on DNA, can be attributed to manifestation of secondary elementary quantization. This, in its turn, can lead to conclusion, that each one of individual organisms requires a personal distribution of microelements.

Living nature becomes personified!

Necessary biological microelements are included in every part of living organisms and in some cases they naturally concentrate in large quantities in certain organs. Barium has a tendency to accumulate in the eye's pigment and the iris, fluorine—on the teeth and bones, arsenic—in parenchymatous organs, radioactive copper—in bone marrow (Vavilova, 1992).

Medical practitioners have long noticed some patterns, some of which have been theoretically proven, while the rest still await for their explanation. In particular, the important role of ions of natrium and potassium in membrane processes is well known. Copper participates in generation of blood and the general processes of growth and development of an organism, it stimulates hormonal activity of the frontal sector of hypophysis. Calcium, magnesium, strontium, barium and radium participate in the forming of bone structures.

Boron plays a significant role in the reproduction function of a human being. Simultaneously, it actively participates in the process of carbohydrate exchange.

Aluminum has an active effect on connective tissues. Thallium, just as boron, actively influences the nerve system. Silicon is important for the supporting and protective functions of connective tissue, while lead and tin—for the functions of the central nerve system. Sulfur is very important for the vital activity, because it is part of some amino acids (cystine, cysteine, metionine), as well as enzymes and hormones, where it participates in vitally important oxidizing-reconstructive processes.

In medical practice, the deficit of endogenous (i.e. produced by the organism) microelement is usually treated by an artificial infusion of such microelement from outside using an outside source (exogenous microelement), being guided by empirical rules: "... the daily required amount of the element X is, approximately, 0.1 mg, however, since the biological availability of the element X is not large, .20 mg should be artificially added each day...". Therefore, with the deficit of one or another microelement, an organism must receive an extra amount of it. The consequences of such "treatment" are quite difficult to predict.

While hyper vitamin-rich conditions are described in detail, information about pathologies (disorders), caused by an excess content of microelements, are poor and contradictive. Meanwhile, it is obvious, that the evolutionary process inevitably leads to optimization of every system. Microelementary content is not an exception, and the content of microelements in a live organism should also be precisely balanced.

With the help of the effect of the transmutation of isotopes, we can solve the problem of prevention and treatment of diseases, caused by deficit of microelements, which would also allow to avoid overloading an organism by exogenous microelements, while preserving the therapeutic effect, due to mobilization of an organism's internal resources. The main importance of such approach becomes evident, if we consider the need of a personified distribution of microelements. No average (throughout the whole kind) parameters will be of any help here.

An optimal solution consists of endogenous compensation for the mentioned deficit of microelements, when an organism receives isotopic precursors (initial components for transmutation) of the required microelements from outside.

Then, a reaction of transmutation takes place inside an organism, producing the necessary microelements. With such method of prevention and treatment of diseases, etiologically caused by deficit of microelements, the quantity of synthesized microelements, which will be processed and stored in an organism, will be exactly the amount, which each individual organism needs, and they will be located only in those parts of an organism, where they are needed. Such self-regulation would be a direct result of the fluctuational process of isotopic synthesis, described in chapter 5, which is induced by the process of nuclear interaction in an optimal quantum system, and which becomes fixed in the process of adaptive accommodation to the structure of a biological object.

6.2 Obtaining rare light and heavy stable isotopes by biotechnological methods, based on the phenomenon of transmutation of isotopes

Producing stable isotopes is one of the most promising areas of technology.

Isotopes are widely used in microelectronics, medicine, pharmacy, study of materials and other fields of science and technology. Nevertheless, the use of this technology is limited at the present moment (especially in the case of rare and very rare isotopes) due to very high costs of the enrichment process and separation of required isotopes from their natural sources—chemical compounds. The methods of transmutation of isotopes in biological systems, examined earlier, make it possible to forecast the possibility of using processes of low temperature nuclear synthesis in growing microbiological cultures for producing rare isotopes, as well as for their optimal use.

The data below contains information about some of the rare isotopes, which can be obtained with the help of transmutation of isotopes in growing microbiological cultures. Numbers in parenthesis, following the name of an isotope, represent its content (in percentage points) in their natural state, as part of a given chemical element

$B^{11} + d^2 = C^{13}(1.0\%)$

$Be^9 + Be^9 = O^{18}(0.2\%)$

$B^{11} + He^4 = N^{15}(0.4\%)$

$F^{19} + N^{14} = S^{33}$ (0.7%)

Al^{27}(100%) $+ d^{2} = Si^{29}$(4.0%)

$P^{31} + Be^{9} = K^{40}$(0,001%)

$P^{31} + Be^{11} = Ca^{42}$(0.6%)

Cl^{35}(75%) $+ Li^{7} = Ca^{42}$(0.6%)

Cl^{37}(25%) $+ Li^{7} = Ca^{44}$(2%)

Ti^{48}(74%) $+ Be^{9} = Fe^{57}$(0.6%)

Co^{59}(100%) $+ d^{2} = Ni^{61}$(0.6%)

Cu^{65}(31%) $+ d^{2} = Zn^{67}$(0.6%)

Il these reactions are barrierless and exothermal and, therefore, can run without any additional energy in the process of biological systems' growth (microbial cultures, for example). The final product of these reactions are isotopes of chemical elements, required for growth of cultures. This, in turn, suggests, that in order to optimize these reactions, chemical elements, which are analogues of generated isotopes, should not be present in the nutrient medium (or be contained in minimal quantities). In this optimal case, the synthesized isotopes will become immediately involved in vital processes of growing cultures, which would prevent them from participating in the process of reverse decomposition into initial isotopic components.

Apart from formation of light nuclei of chemical elements and isotopes up to nuclei of the iron group on the basis of synthesis reactions, the effect of transmutation of isotopes makes possible synthesis of heavier nuclei $_{Z}X^{A}$ up to transuranium ones.

Possibility of running energy efficient reactions of synthesis of heavy nuclei is determined by the requirement of minimization of energy of the system of interacting nuclei.

It follows from the Bete-Waitzzekker formula, that the full energy of a nucleus can be represented by the expression

$$E(Z,A) = (A-Z)M_{n}c^{2} + ZM_{p}c^{2} - \varepsilon_{1}A + \varepsilon_{2}A^{2/3} + \varepsilon_{3}Z^{2}/A^{1/3} + \varepsilon_{4}(2A-Z)^{2}/A - \delta\varepsilon_{5}/A^{1/2}$$

Here, $(A-Z)M_{n}c^{2}$ and $ZM_{p}c^{2}$ is the sum of rest energies of all $(A - Z)$ neutrons and Z protons of the nucleus; $\varepsilon_{1}A$—the energy of links of all nucleons in a nucleus (without other characteristics, determined by other components of the Coulomb interaction of nucleons, surface effects and effects of coupling of nucleons), $\varepsilon_{1} \approx 15{,}56$ MeV; $\varepsilon_{2}A^{2/3}$—

surface energy of the nucleus ($\varepsilon_2 \approx$ 17,23 MeV); $\varepsilon_3 Z^2/A^{1/3}$—Coulomb energy of mutual repulsion of protons in the nucleus ($\varepsilon_3 \approx$ 0,697 MeV); $\varepsilon_4 (2A\text{-}Z)^2/A$—"symmetric" energy of the system of nucleons, which increases with violation of parity of the number of protons and neutrons due to necessity of occupying higher energy levels because of the Pauli principle for these fermions ($\varepsilon_2 \approx$ 23,28 MeV); ($\delta\varepsilon_5/A^{1/2}$—energy of coupled correlations between identical nucleons, provided that $\delta(A, Z) = 0$ for nuclei with odd value of A, $\delta(A, Z) = -1$ for even A and even Z, $\delta(A, Z) = 1$ for even A and odd Z, $\varepsilon_5 \approx$ 12 MeV.

The condition of energy efficiency of reaction of synthesis of heavy nuclei

$$_{Z1}X^{A1} + {}_{Z2}X^{A2} = {}_{Z3}X^{A3},$$

running with altering the charge $Z_3 = Z_1 + Z_2$ and mass $A_3 = A_1 + A_2$ of initial isotopic components, is determined from the condition of positive change of energy of links of the nuclei, participating in the reaction

$$\Delta E = E(A_1Z_1) + E(A_2,Z_2) - E(A_3,Z_3) \geq 0$$

The physical justification of energy efficiency of such synthesis reactions with participation of heavy and super heavy nuclei is the fact, that with interaction of a heavy or a super heavy nucleus with a very light isotope (from hydrogen to carbon), increase of the linking energy, related to absorption of the light isotope, is more significant than decrease of such energy due to an increase of the mass of a heavy or a super heavy nucleus.

For example, a positive energy yield in the synthesis reaction with technetium will be

$$Th^{232} + He^3 = U^{235},\ \Delta E = 9{,}6\ \text{MeV}$$

and isotopes of rhenium

$$Re^{185} + p^1 = Os^{186}\quad \Delta E = 6{,}5\ \text{MeV};$$

$$Re^{185} + d^2 = Os^{187}(1.6\%),\ \Delta E = 6{,}8\ \text{MeV};$$

$$Re^{187} + p^1 = Os^{188},\ \Delta E = 7{,}3\ \text{MeV}$$

Energy efficiency of the last of the examined reactions demonstrates, in particular, that in natural conditions all rhenium isotopes gradually transform into isotopes of osmium during the process of transmutation in biological systems.

Let's consider, as an example, a list of several other reactions of synthesis of the nuclei (including those, that are considered rare, because they are present in their respective chemical elements in very small quantities), which are heavier than iron and are characterized by positive energy yields

$$J^{127} + Li^{7} = Ba^{134}\ (2.4\%)$$

$$Ta^{181} + Li^{6} = Os^{187}\ (1.6\%)$$

$$Ta^{181} + Li^{7} = Os^{188}$$

$$W^{183} + D^{2} = Re^{185}.$$

Rhenium is the element, which in biological reactions replaces manganese.

We can also propose the reaction of transmutation

$$Os^{190} + Li^{7} = Au^{197},$$

which, throughout the ages, has been the "golden dream" of all generations of alchemists and which naturally leads to synthesis of gold in the volume of growing microbiological culture. It follows from the data provided in chapter 5, that potassium (the ionic radius of which is equal to 1.33 A with the charge of the ion $Z^* = +1$) is the stereochemical analogue of gold (with the ionic radius of 1.37A with the same charge $Z^* = +1$). In the process of culture's growth, potassium required for that process would be substituted by gold. This reaction will be successful only if the nutrient medium, where a bacterial culture grows, is poor in potassium and has an excess quantity of osmium and lithium. As follows from calculation of mass defects, this reaction is exothermal and barrierless, it doesn't require any energy inputs, while it will yield energy $\Delta E = 7.37$ MeV.

Of course, producing such reaction takes certain effort and knowledge, since it would require selection of such soluble salts of osmium and lithium, which wouldn't inhibit the culture's growth by themselves and, beside that, the culture itself must be adapt to synthesized gold (which also shouldn't inhibit its growth).

But this is the question of professional microbiology.

Platinum is no less valuable metal. It is possible to give several examples of exothermal barrierless reactions, which would result in synthesis of two different isotopes of platinum

$$Re^{187} + Li^{7} = Pt^{194},\ \Delta E = 8.47\ \text{MeV};$$

$$W^{186} + Be^{9} = Pt^{195},\ \Delta E = 1.65\ \text{MeV}.$$

These reactions would be possible if the nutrient medium, in which a bacterial culture grows, is poor in iron and has an excess quantity of rhenium and lithium (in the first case) or tungsten and beryllium (in the second). Since ionic radiuses of double-valency iron (0.75–0.83 A) and double-valency platinum (0.8–0.85 A) are practically identical, in the absence of iron synthesized atoms of platinum will be accepted by a growing microbial culture.

6.3 On the possibility of radioactive waste utilization using the effect of transmutation of isotopes in biological systems

Beside generating heavier stable isotopes from light, medium and heavy stable isotopes on the basis of synthesis reactions, there is also possibility of utilization of light, medium and heavy radioactive isotopes (for example, components of spent nuclear fuel or isotopes used in metrology) by the way of their transformation into stable isotopes of chemical elements, which are necessary for the vital activity of microbial cultures.

The problem of utilization of such wastes is very important for a number of reasons. The overall share of nuclear energy in global energy production is, according to various estimates, from 10% to 20%. In some industrialized countries, atomic energetic is predominant—its share reaches 70% from the total amount of produced and consumed energy resources. However, the development of atomic sources of energy is currently all but frozen. The main reason is not the shock of Chernobyl disaster, but the lack of an acceptable solution for processing spent nuclear fuel, as well as storage and utilization of radioactive waste. Let's examine this problem more closely.

It is known, that power reactors can be of two main types: reactors on fast neutrons, using highly enriched uranium, and reactors on thermal neutrons, working on uranium with the level of enrichment of 3–15% of U^{235} (depending on the type of the reactor—light water, heavy water, uraniumgraphite channel, gas, etc.).

In the process of a reactor's operation, splitting U^{235} results in generation of several isotopes having a very large capacity for capturing thermal neutrons (in particular, isotopes of xenon), which results in continuous decrease of reactor's radioactivity, as burning-out increases (the effect of so called "xenon poisoning"). It can be alleviated, to a certain level, by regulating rods of the reactor's controlling system. After reaching a certain degree of burning-out, reactor becomes, essentially, uncontrollable. Depending on the type of a reactor, working on thermal neutrons, an admissible degree of burning-out is from 5 to 20%, after which blocks of heat-producing elements (HPE) must be dismantled and replaced. Therefore, at least 80% of expensive enriched U^{235} in reactors, working on thermal neutrons, remains unused.

Situation is somewhat better for reactors, working on fast neutrons, because the capacity for capturing fast neutrons for various isotopes differ much less, than for thermal neutrons. The outcome for such a reactor is a significantly lower risk of xenon poisoning (or other kind of deficiency). However, due to the fact, that in the process of splitting uranium nucleus, 2–3 nuclei of other elements are formed, the density of nuclear fuel constantly decreases as it is irradiated by neutrons, while its volume increases. This effect is called "radioactive swelling" of the HPE's, resulting in the danger of impossibility of their removal from reactor in case of its overloading. Because of that, permissible burning-out for reactors, working on fast neutrons, is also limited and usually doesn't exceed 40%.

In the end, reactors on fast neutrons use less than half of their nuclear fuel. It is necessary to note, that these reactors use fuel with a higher degree of enrichment by isotope U^{235}, which makes it more expensive.

In all countries, developing their own nuclear energy, attempts have been made over several decades to process irradiated fuel for the purpose of regeneration of unspent enriched U^{235}. Both classical methods of hydrometallurgy using sorption and extraction processes, which have proven themselves worthy at initial redistribution and affinage of natural uranium, and quite exotic piro-metallurgical (for example, using melted salts) and fluoride technologies, promising good results, were studied.

All these promising, at first, studies and developments did not bring any concrete results and were stopped everywhere. Few existing technological lines for treating irradiated nuclear fuel on the basis of low-enriched uranium (HPE's of reactors on thermal neutrons) operate using classical extraction and sorption processes, similar to the ones used in affinage of natural uranium.

The fate of reactors on fast neutrons is yet unclear. Attempts to develop a satisfactory technology for processing irradiated fuel of such reactors were un successful and currently that fuel is not processed at all. Apparently, existing and most common, at present, light water water-based and uranium-graphite reactors will remain, in the foreseeable future, the main types of energy-producing nuclear reactors.

Each year, about 10 thousand tons of spent nuclear fuel (SNF) is unloaded from reactors of different types. At first, almost all countries, developing their own nuclear power, made attempts to process irradiated fuel, but after many years and repeated unsuccessful attempts to develop economically viable and ecologically sensitive technology, abandoned the idea of processing used nuclear fuel. At the present time, such facilities still exist in France, Great Britain and Russia, although both in France and in Great Britain, the possibility of their closing is being discussed. Capacity of those facilities in three countries allow to process only about 10% of SNF, unloaded annually from reactors of nuclear power plants (NPP) in those countries. The United States has principally declined any attempts to process SNF since the 70's in favor of storing whole blocks of spent HPE's in special storage sites.

There is over 100 thousand tons of unprocessed SNF in the world today.

Existing storage capacities are close to becoming full. Therefore, keeping the status quo with respect to the present situation with processing and accumulation of SNF will, sooner or later, inevitably lead to either the need to reject the use of atomic energy, or, at least, to a significant reduction in construction of new NPP's in most densely populated (i.e. most industrially developed) countries.

Paradoxically, the main advantage of NPP's in comparison with other power generating means is possibility of their construction in close vicinity to main consumers of electric and thermal power, near big urban centers. Loosing that advantage as a result of forced relocation of NPP's as well as SNF storage and processing facilities away from main consumers, will jeopardize the use of atomic energy in general.

There are two reasons why almost all countries have rejected the idea of processing SNF.

The first one is purely economical. Initially, processing of SNF was conducted with the purpose of extracting unburnt fission material (enriched uranium) with subsequent refurbishing of HPE's from regenerated uranium.

Costs of chemical processing were considered to be much lower, than costs of separating isotopes of uranium, i.e. its enrichment. However, as the technology of isotopic separation moved forward, and, especially, with overall turning from gas-diffusion to centrifugal separation, the costs of uranium enrichment became much lower. On the other side, over many years, all attempts to develop a technology for purification of secondary uranium from products of nuclear fission, providing a high degree of background activity of regenerated uranium, have been unsuccessful.

Combination of these factors has resulted in a situation, when economic benefits of SNF processing with the purpose of obtaining unburnt fission material was lost. The limited quantity of discovered reserves of natural uranium and the possibility of their eventual exhaustion remained hc only justification of the need to develop technologies for SNF processing.

The second reason is less obvious, but more serious. Existing technologies for SNF processing are not waste-free and operating facilities do not use closed-cycle technologies. Moreover, as a result of SNF processing, the volume of radioactive waste (in mass equivalent), which can not be further processed using contemporary technologies and is subject to burial at special storage sites, has increased by hundreds and thousands of times in comparison with their initial amount (naturally, with correspondingly lower level of their nominal activity).

For example, one ton of SNF produces about the same or slightly larger quantity of solid and very highly radioactive waste (concentrated radioactive isotopes of various elements—products of fission, encapsulated in glass or concrete), which must be stored in special burial sites for highly radioactive waste for no less than a thousand years. Beside that, 10–20 tons of solid and liquid waste of medium activity is generated (a mixture of products of fission with various non-active substances and materials used in the process of separation and processing of HPE's), which also need to be buried.

Finally, there are several hundreds of tons of gaseous, liquid and solid low radioactive waste: various gaseous exhausts, flushing waters, drains, dismantled equipment, spent extracting agents and absorbents, contaminated uniforms and personal means of protection and other waste associated with any chemical technology, contaminated by radioactive materials. These wastes need to be stored in regular burial sites, not requiring special expensive equipment. However, their volume, in contrast with highly radioactive wastes, is huge.

Therefore, as a result of processing of unloaded SNF using the existing technologies, 10 million tons of low radioactive waste can be generated each year. It doesn't account for gaseous wastes, which are simply discharged into the atmosphere after their dissolving with air to a relatively safe concentration of radioactive contaminants. Note, that the overall dose of such substances remains the same.

Taking these factors into consideration, we can justifiably conclude, that storing SNF is cheaper and safer than their processing using the existing technologies.

Despite extremely high nominal expenses required for the storage of unprocessed SNF, tremendous costs of building and maintaining of specialized storage facilities for highly active radiologically dangerous wastes, it is still much cheaper and safer, by the opinion of many experts, than processing of radioactive waste using today's technologies. However, as was mentioned earlier, existing storage capacities for SNF around the world are close to being filled up, while construction of new ones in densely populated areas and countries (i.e. in places of highest demand) is practically impossible.

Limited capabilities of existing chemical technologies are due to the fact, that, from the point of view of classical chemistry, regenerated uranium, as well as different low radioactive wastes (solid, liquid and gaseous) are chemically pure: the content of radioactive isotopes as chemical admixtures is very small. These wastes are considered "dirty" only from the point of view of radiochemistry. At present, there are no effective chemical, physical-chemical or physical processes and technologies, capable of removing such insignificant quantities of radioactive contaminants at reasonable costs.

After a three-year holding in cooling ponds at NPP's, total activity of HPE's is reduced by, approximately, ten times, due to decay of most short-lived isotopes.

From remaining activity, 99% is contributed by relatively long-living isotopes of strontium, zirconium, ruthenium, cesium, iodine and rare earth elements. Some activity is also created by isotopes of noble gases.

Chemical processes are effective in removing rare earth elements, iodine, noble gases and a significant portion of strontium and cesium from regenerated uranium. However, removal of ruthenium isotopes, which, due to its chemical characteristics (numerous valency states and easiness of shifting them) is distributed throughout the entire technological chain, contaminating it in the process, presents a serious challenge.

Low radioactive waste is contaminated by radioactive isotopes of virtually the same elements, with the exception of isotopes of xenon, krypton, and iodine.

Activity of gaseous waste is determined (apart from aerosols) mainly by isotopes of noble gases (krypton and xenon) and iodine, generated primarily in the result of decay of xenon isotopes.

The effect of transmutation of isotopes in bacterial cultures, stable in extremely high levels and doses of ionizing radiation, provides hope, that solutions to these problems may be found in the use of biological systems.

In contrast with classical chemical processes, growing micro-biological systems can provide an extremely high selectiveness and completeness of extraction of various chemical elements (including isotopes) from different media, and the phenomenon of low temperature transmutation can allow transformation radioactive isotopes, absorbed by a bacterial culture, into stable isotopes of other chemical elements. Naturally, we are not talking (for obvious reasons) about a full-scale implementation of the use of microbiological systems in the initial stages of SNF processing and processing of highly radioactive waste.

The prospects of using bacterial cultures are, to our opinion, the following:

1. Fine purification of regenerated uranium on the final stages of its separation from products of decay to a level, allowing it to be used for refabrication of HPE's without using heavy chambers and remote controls.
2. Purification of gaseous exhausts of NPP's and other facilities, processing SNF, from isotopes of noble gases and products of their decay (mainly, iodine).
3. Purification of low radioactive waste to a level, providing a possibility of their burial, as non-radioactive waste, i.e. to the level of natural radioactive background (0.8-2).10^{-5} Roentgen/h.
4. Purification of sewage and drain waters of NPP's and other facilities, processing SNF to the level of natural background.

Apart from using processes of transmutation (and microbiological systems in general) in purification procedures of various substances and materials from radioactive waste, another potential use of transmutation in production or selective extraction of certain radioisotopes with a very high degree of radiochemical (and chemical) purity—to be used in medicine, for example—such as isotopes of technetium, gallium, iodine and others, production of which is currently quite expensive due to the need of high degree of radiochemical purity in some radioisotopic materials, also appears quite promising. We believe, that using biological systems, the costs of making radiopharmacological materials with a required degree of radiochemical purity, could be substantially lowered.

No less promising, is possibility of using biological systems for solving specific problem of decontaminating cooling surfaces of nuclear reactors. In the process of a reactor's operation, because of leaching of constructive materials in the water environment, under conditions of strong neutron fields, a significant amount (from the radiochemical perspective) of active cobalt is produced. This element tends to form compact depositions outside the zone of biological protection of a reactor. As a result, in order to prevent additional exposure by maintenance personnel, a larger biological protection zone—up to the second contour, inclusive—is needed for such reactor. In the 80's, this problem was, in general, successfully resolved through adding zinc compounds to the contents of water, circulating in the cooling contours. With the presence of zinc, cobalt didn't form compact deposits, since it remained in the cooling liquid in the form of a colloid solution. However, one of isotopes of zinc—Zn^{64}, which natural content in ordinary natural zinc is about 50% becomes even more active because of neutron radiation, than radioisotopes of cobalt. Therefore, instead of natural zinc, so-called "depleted zinc", not containing isotope Zn^{64}, has to be added to cooling liquid. "Depleted zinc" is quite an expensive product, as all separated stable isotopes. Beside that, adding zinc causes, in its turn, a series of other problems, related to cleaning reactor's contours. Considering a relatively low level of ionizing radiation in the first cooling contour outside the active zone of a reactor, and still lower one in the second contour, possibility of applying biological systems in general and processes of low temperature transmutation in particular for removing active cobalt from the contours (or, at least, prevent generation of its compact deposits) appears to offer a lot of potential.

Let's consider in more detail the possible ways of using the effect of transmutation of radioactive isotopes in microbiological systems.

During the reaction of splitting isotopes of uranium U^{235} (half-life—$7 \cdot 10^8$ years) in nuclear reactors, large number of radioactive isotopes (highly active waste) is generated with medium and long half-life periods:

Sr^{90} (half-life τ= 28.5 years, relative activity of waste (in relation to the activity of Pu^{239}) immediately upon extraction from a reactor is equal Q = 230),

Zr^{95} (τ=64 days, Q = 5800),

Nb^{95} (τ=35 days, Q = 5700),

Mo^{99} (τ=66 hours, Q = 6100),

Ru^{103} (τ=39 days, Q = 3800),

Ru^{106} (τ=373 days, Q = 860),

Sb^{125} (τ=2.8 years, Q = 150),

I^{131} (τ=8 days, Q = 3100),

Cs^{134} (τ=2 years, Q = 170),

Cs^{137} (τ=30.2 years, Q = 260),

Ce^{144} (τ=285 days, Q = 3900),

Eu^{154} (τ=8.8 years, Q = 14),

Pu^{238} (τ=87.7 years, Q = 1.3),

Pu^{239} (τ=24000 years, Q = 1),

Pu^{240} (τ=6550 years, Q = 1.5),

Pu^{241} (τ=14.4 years, Q = 180),

Am^{241} (τ=432 years, Q = 0.16).

Approximate accumulation of major long- and medium living products of fission and transuranium elements in the irradiated fuel of power reactors at average burning-out of 11 MW per day, counting per 1 kg of uranium, is presented in Table 6.1.

The spectrum and data concerning products of fission are provided for reactors on thermal neutrons at average burning-out of fuel; for reactors on fast neutrons, the spectrum of fission products is slightly shifted towards a higher share of longer-living radionuclides and transuranium elements. Table 6.1 contains information only about the main elements of fission products, comprising over 99.9% of residual total activity of irradiated fuel after 2–3 years of holding in cooling ponds. The same table provides information about stereochemical analogues of radioactive nuclei, which substitute them (or substituted by them) in any combination.

One of the most effective methods of radioactive waste utilization consists of separating unspent nuclear fuel from such waste for its further use.

Most problems, associated with utilization of highly radioactive waste and its subsequent use, stem from presence of highly active isotopes Zr^{95}, Nb^{95}, Ru^{103}, Ru^{106} with relatively short half-life periods (in comparison with the main “fuel” isotopes U^{235}, Pu^{240}, Pu^{239}, which may be used again as heat-producing elements for NPP’s after their regeneration). These isotopes constitute a relatively small portion of spent nuclear fuel’s mass, but they contribute significantly to its activity. Irradiation, produced by them, is the factor, preventing regeneration of spent fuel and extraction of “regular fuel isotopes” from it (U^{235} and all plutonium isotopes), suitable for repeated use in reactors.

Table 6.1

Radionuclide	*Activity, Bq*	*Charge and radius of ion*	*Stereochemical analogues*
Sr^{90}	2-4.10^{17}	Z=+2, R=1.2±0.07A	Ca^{2+} Eu^{2+} Hg^{2+} Sm^{2+}
Zr^{95}	5-9.10^{18}	Z=+4, R=0.8±0.03A	Hf^{4+} La^{4+} Nb^{4+}
Nb^{95}	5-9.10^{18}	Z=+4, R=0.72±0.05A Z=+5, R=0.68±0.02A	Mo^{4+} Os^{4+} Se^{4+} Sn^{4+} W^{4+} Sb^{5+} Bi^{5+} Ta^{5+}
Mo^{99}	5-9.10^{18}	Z=+4, R=0.67±0.01A Z=+6, R=0.66±0.03A	Ir^{4+} Nb^{4+} Pd^{4+} Pt^{4+} W^{6+}
Ru^{103}	4-8.10^{18}	Z=+3, R=0.77A Z=+4, R=0.63±0.02A	Ir^{3+} Mn^{3+} Rh^{3+} Ir^{4+} Mo^{4+} Nb^{4+} Os^{4+} Tl^{4+} W^{4+}
Ru^{106}	1-2.10^{18}	Z=+3, R=0.77A Z=+4, R=0.63±0.02A	Ir^{3+} Mn^{3+} Rh^{3+} Ir^{4+} Mo^{4+} Nb^{4+} Os^{4+} Tl^{4+} W^{4+}
Sb^{125}	1-3.10^{16}	Z=+3, R=0.9A Z=+5, R=0.62A	Au^{3+} In^{3+} Lu^{3+} Ta^{5+}
I^{131}	3-5.10^{18}	Z=+5, R=0.94A Z=-1, R=2.2±0.02A	Pa^{5+} U^{5+} SH^{-1} MSO_4^{-1}
Cs^{134}	1-4.10^{17}	Z=+1, R=1.65±0.02A	Fr^{1+} K^{1+} NH_4^{1+} Rb^{1+}
Cs^{137}	3-5.10^{17}	Z=+1, R=1.65±0.02A	Fr^{1+} K^{1+} NH_4^{1+} Rb^{1+}
Ce^{144}	3-6.10^{18}	Z=+3, R=1.1±0.08A Z=+4, R=0.95±0.07A	Bi^{3+} Gd^{3+} La^{3+} Nd^{3+} Pa^{3+} Pr^{3+} Th^{3+} La^{4+} Pa^{4+} Te^{4+}
Eu^{154}	1-3.10^{16}	Z=+2, R=1.19A Z=+3, R=1.07±0.1A	Hg^{2+} Pb^{2+} Yb^{2+} Bi^{3+} Ce^{3+} Dy^{3+} Nd^{3+} Pa^{3+} Sm^{3+} Tb^{3+} Th^{3+} Tm^{3+} Tl^{3+} Y^{3+} Yb^{3+}
Pu^{238}	1-2.10^{15}	Z=+3, R=1.02±0.01A Z=+4, R=0.88±0.02A Z=+5, R=0.87A	Dy^{3+} Ho^{3+} Nd^{3+} Pa^{3+} Tb^{3+} Tm^{3+} U^{3+} Y^{3+} Yb^{3+} La^{4+} Pa^{4+} Te^{4+} Zr^{4+} Pa^{5+}
Pu^{239}	1-2.10^{15}	Z=+3, R=1.02±0.01A Z=+4, R=0.88±0.02A Z=+5, R=0.87A	Dy^{3+} Ho^{3+} Nd^{3+} Pa^{3+} Tb^{3+} Tm^{3+} U^{3+} Y^{3+} Yb^{3+} La^{4+} Pa^{4+} Te^{4+} Zr^{4+} La^{4+} Pa^{4+} Te^{4+} Zr^{4+} Pa^{5+}
Pu^{240}	1-2.10^{15}	Z=+3, R=1.02±0.01A Z=+4, R=0.88±0.02A Z=+5, R=0.87A	Dy^{3+} Ho^{3+} Nd^{3+} Pa^{3+} Tb^{3+} Tm^{3+} U^{3+} Y^{3+} Yb^{3+} La^{4+} Pa^{4+} Te^{4+} Zr^{4+} Pa^{5+}
Pu^{241}	1-2.10^{17}	Z=+3, R=1.02±0.01A Z=+4, R=0.88±0.02A Z=+5, R=0.87A	Dy^{3+} Ho^{3+} Nd^{3+} Pa^{3+} Tb^{3+} Tm^{3+} U^{3+} Y^{3+} Yb^{3+} La^{4+} Pa^{4+} Te^{4+} Zr^{4+} Pa^{5+}
Am^{241}	1-2.10^{14}	Z=+3, R=1.0±0.01A Z=+4, R=0.87±0.02A	Dy^{3+} Ho^{3+} Nd^{3+} Pa^{3+} Tb^{3+} Tm^{3+} U^{3+} Y^{3+} Yb^{3+} La^{4+} Pa^{4+} Te^{4+} Zr^{4+}
Am^{244}	2-4.10^{14}	Z=+3, R=1.0±0.01A Z=+4, R=0.87±0.02A	Dy^{3+} Ho^{3+} Nd^{3+} Pa^{3+} Tb^{3+} Tm^{3+} U^{3+} Y^{3+} Yb^{3+} La^{4+} Pa^{4+} Te^{4+} Zr^{4+}

The process of extraction and separation of certain kinds of highly active isotopes of low concentration from multi-component radioactive waste during the process of growth of maximally radiation-resistant microbiological cultures like *Deinococcus Radiodurans* (capable of sustaining normal metabolism at radiation levels of over 1–10 Mrad), as well as cultures, capable of withstanding considerably lower doses, but nevertheless reaching 30 Krad and more) is based on the fact, that the chemical characteristics of different isotopes of the same element and their ability to participate in vital processes are practically indiscernible.

Because of that, a growing culture can use radioactive isotopes, present in the nutrient medium, for sustaining its vital activity (the process of metabolism).

Moreover, there is information suggesting an important role of radioactive elements in metabolism of live objects. Such hypothesis was offered, for the first time, by Vernadski, which was later corroborated by modern studies.

Thus, in the work of Sobotovich (Sobotovich, 1994), the ability of micromyces to use such seemingly inert substratum as highly radioactive reactor-grade graphite in constructive metabolism, was convincingly demonstrated. The use of uranium as the donor of electrons in energetic metabolism of *Thiobacillus ferrooxidans* has been experimentally proven (Soljanto, 1980). This proves, beside other things, the importance of selecting a specific microelementary content of an environment (madium) for the process of sustaining normal vital activity of biological systems.

Let's examine possible schemes of extracting highly active admixtures in the process of their selective transmutation into stable isotopes in growing radiation-resistant biological cultures.

Since all examined highly active wastes are medium (by their mass numbers) isotopes, the process of their transmutation (a synthesis-like reaction)is only possible with such a pattern of nuclear transformations, for which an element, being synthesized, immediately used by a growing culture in the process of its metabolism, would be heavier, than an isotope, being utilized.

Based on these principles, the following possible reactions of transmutation of radioactive waste may be proposed.

6.3.1 *Reaction of utilization of long-living isotope* Sr^{90}

Enrichment of standard nutrient medium, necessary for growth of a microbiological culture, by salts containing boron, leads to a possibility of running an energy efficient reaction of transmutation of strontium, causing a growing culture to accept the generated unstable short-living isotope of molybdenum

$$Sr^{90} + B^{9} = Mo^{99}$$

After getting involved in the metabolism process of a growing microbiological culture, this isotope decays within 66 hours. In this utilization scheme, a radiation-resistant culture performs the function of providing accelerated deactivation of a specific long-living isotope by the way of its transformation into a short-living isotope.

6.3.2 *Reaction of utilization of radioisotope* Zr^{95}

This reaction can be run at the absence of helium in the nutrient medium (for example, due to its forced blowing out) and leads to an energy efficient reaction

$$Zr^{95} + He^{4} = Mo^{99}$$

with generation of the unstable, short-living isotope of molybdenum and its acceptance by a growing culture. This isotope has the half-life period of 66 hours and decays after its acceptance by a growing culture.

6.3.3 *Reaction of utilization of* Nb^{95}

In order to produce the reaction of utilization of highly active isotope Nb^{95}

$$Nb^{95} + p^{1} = Mo^{96}$$

it would be sufficient to have hydrogen atoms in the nutrient medium (along with other necessary components), which is always the case, since hydrogen is one of the vitally important chemical elements. The result of the reaction is the stable isotope Mo96, involved in the metabolism process of a microbiological culture.

6.3.4 *Reactions of utilization of highly active isotopes* Ru^{103} *and* Ru^{106}

Reactions of utilization of isotopes Ru^{103} and Ru^{106}may be accompanied by generation of very rare isotopes of tin Sn^{115} and Sn^{117} (natural concentration is 0.4% and 2.4%, respectively)

$$Ru^{103} + C^{12} = Sn^{115}\ (0,4\%)$$

$$Ru^{106} + C^{12} = Sn^{118}\ (2,4\%)$$

and require presence of carbon in the nutrient medium, which is always the case, since carbon is one of the vitally important chemical elements, necessary for culture growth, and, therefore, is always present in the nutrient medium. The utilization scheme of this radioisotope is similar to that for Sr^{90}.

6.3.5 *Reactions of utilization of highly active nuclei of* Sb^{125}

Reactions of utilization of radioactive Sb^{125}isotope

$Sb^{125} + B^{10} = Ba^{135}$ (6,6 %)

$Sb^{125} + B^{11} = Ba^{136}$ (8 %)

may run during the growth of a radiation-resistant culture or microbiological associations and with the presence of boron salts in the nutrient medium and absence of barium salts.

6.3.6 *Reactions of utilization of highly active isotope* J^{131}

Reactions used for utilization of this isotope

$J^{131} + Li^{6} = Ba^{137}$ (11%)

$J^{131} + Li^{7} = Ba^{138}$ (71,7%)

may occur in a radiation-resistant culture such as Deinococcus Radiodurans, based on the nutrient medium, containing, beside regular components, isotope J^{131}—the object of utilization—and lithium salt.

In addition to utilization of radioisotopes, contained in spent nuclear fuel, the method also allows utilization of spent highly active sources of isotopes, usually used in metrology and sterilization by radiation.

Highly active sources of hard gamma-radiation are widely used industrially in metrology, flaw detection and geology. After expiration of their operational resource (due to reduction of activity because of deteriorating characteristics of their protective systems), these sources become unsuitable for further use.

Because of their high costs, it does not make sense economically to reactivate them by separating isotopes from the protective system with further radiochemical cleaning and enrichment. Burying these highly active sources also involves high costs and risks of environmental damages. Same problems arise after expiration of operational resource in highly active sources, used in sterilization technology.

Let's consider some reactions, which would allow to utilize such isotopes with the help of the method of transmutation in growing microbiological cultures with subsequent transformation of radioactive isotopes into stable ones (including rare).

The process of transmutation should result in generation of final, heavier than the interacting isotopes, nucleus. Because of that, the maximal atomic number of a transmutated isotope is determined (with simultaneous fulfillment of the law of preservation of energy, which is expressed in increasing mass defects of interacting nuclei) by chemical elements, which are required for a culture's growth and have the highest number and atomic weight. Out of such elements, Cs and Rb (biochemical analogues of potassium) are most important, along with Sr, which, by its chemical characteristics is analogue of calcium. Moreover, when choosing the best suitable reaction, one can be guided by the recommendation, that the final isotope is stable, with the shortest possible chain of intermediate reactions leading towards the generation of the final stable nucleus.

Having those requirements in mind, below is the list of principal highly active sources of isotopes, which may be utilized with the help of the phenomenon of transmutation of isotopes in microbiological cultures:

Co^{60} (half-life—5.2 years), Ar^{39} (half-life—270 years),

Sc^{46} (half-life—18.7 years), Kr^{85} (half-life—10.7 years),

Mn^{54} (half-life—312 days), S^{35} (half-life—87 days)

6.3.7 *Reactions of utilization of highly active isotope* Co^{60}, *used in metrology and flaw detection*

Half-life of this isotope is equal to 5.2 years and it is the main source of gamma-radiation, used in sterilization systems, radiography and metrology.

Reactions of transmutation of this isotope, providing its utilization, may run according to different patterns.

At the presence of nickel in the nutrient medium or presence of gaseous neon in the growth area and the absence of potassium salts, an energy efficient reaction of synthesis may run according to the following schemes:

$$Co^{60} + Ne^{20} = Rb^{80} \text{ (half-life—30 seconds)} \rightarrow Kr^{80} \text{ (2.2\%)}$$

$$Co^{60} + Ne^{22} = Rb^{82} \text{ (half-life—6.3 hours)} \rightarrow Kr^{82} \text{ (11.6\%)}$$

In the presence of sodium in the nutrient medium and the absence of calcium salts, an energy-efficient reaction of synthesis may run according to the scheme

$$Co^{60} + Na^{23} = Sr^{83} \text{ (half-life—32 hours)} \rightarrow Rb^{83} \text{ (86 days)} \rightarrow Kr^{83} \text{ (11.5\%)}$$

In this reaction, there is synthesis of strontium nuclei, which are similar to calcium by their biochemical characteristics and substitute it during a culture's growth.

6.3.8 *Utilization of highly active isotope* Kr^{85}

This isotope is closest to rubidium and strontium and can be transformed in a growing culture in the process of the following reactions.

If the nutrient medium doesn't contain calcium, reactions of transmutation can run with the presence of helium in the area of growth

$$Kr^{85} + He^{4} = Sr^{89} \text{ (half-life—50 days)} \rightarrow Y^{89}$$

If we use a nutrient medium, which is poor in potassium, one of the possible reactions of transmutation with participation of protons of water has the form

$$Kr^{85} + p^{1} = Rb^{86} \text{ (half-life—18.7 days)} \rightarrow Sr^{86} \text{ (10\%)}$$

Another variant of this reaction requires presence of deuterium in the nutrient medium (or, preferably, preparation of liquid nutrient medium on the basis of heavy water)

$$Kr^{85} + d^{2} = Rb^{87} \text{ (28\%)},$$

while the reaction itself is radiation-free, not producing the accompanying ionizing radiation).

6.3.9 *Utilization of highly active isotope* Sc^{46}

This isotope has a relatively small charge and mass number and, therefore, can be transmutated using several alternative reactions.

In case the nutrient medium doesn't contain iron, reactions of transmutation can run in the presence of boron salts, which, in its natural form, consists of two isotopes B^{10} and B^{11}

$$Sc^{46} + B^{10} = Fe^{56} \text{ (92\%)}$$

$$Sc^{46} + B^{11} = Fe^{57} \text{ (2,2\%)}$$

The second one of these reactions is very important, because, apart from accomplishing the main task of utilization of highly active isotope Sc^{46}, it also results in generation of a rare Mossbauer isotope Fe^{57}, frequently used in many areas of science and technology.

6.3.10 *Utilization of highly active isotope* Mn^{54}

This isotope is the closest neighbor of iron, which is one of the vitally important elements and, because of that, can be most effectively transformed in reactions

$$Mn^{54} + p^{1} = Fe^{55} \text{ (half-life—2 years)} \rightarrow Mn^{55},$$

$$Mn^{54} + d^{2} = Fe^{56} \text{ (92\%)},$$

in the nutrient medium based on light or heavy water, deprived of iron.

6.3.11 *Utilization of highly active isotope* Ar^{39}

Utilization of this isotope can be most efficiently achieved in the reaction of transmutation

$$Ar^{39} + d^{2} = K^{41} \text{ (6,7\%)}$$

In a liquid nutrient medium on the basis of heavy water, deprived of potassium.

The result of this reaction is generation of a rare isotope K^{41}, the natural content of which in potassium is small and equal to 6.7%.

6.3.12 *Utilization of highly active isotope* S^{35}

The reaction of utilization

$$S^{35} + Li^{7} = K^{42} \text{ (half-life—12 hours)} \rightarrow Ca^{42} \text{ (0.6\%)}$$

can be performed in a nutrient medium, deprived of potassium, but with added salts of lithium.

This reaction results in generation of a very rare stable isotope Ca^{42}.

In addition to examined processes of utilization, we should also note one aspect.

In a series of conducted experiments, studying the phenomenon of transformation of isotopes in growing microbiological cultures, we have discovered, that adding one of the cations Li^{+}, Na^{+}, K^{+}, Rb^{+}, Cs^{+}, Tl^{+} to a nutrient medium with a growing biological culture, results in a significant acceleration (by approximately 1.8–2 times) of the reaction of transmutation of other elements and isotopes, not associated with these elements by schemes of nuclear reactions. This result was mentioned earlier, in chapter 3. Given our present level of knowledge, we can not offer an exhaustive explanation of the observed phenomenon. Supposedly, the named single-valency cations serve as catalysts, in the manner, similar to the effect of ionizing radiation ("wave catalysis").

Apparently, this effect can also be used for optimization of radioactive isotopes' utilization.

6.4 Conclusion

The examples, provided above show, that the effect of low temperature transmutation of isotopes may be used in solving a very wide array of tasks—from issues of life support and forming a new perspective on human pathogenesis to the problem of utilization of spent nuclear fuel.

Consistent application of the ideology of transmutation of isotopes and chemical elements in living organisms may lead to reevaluation of a wide specter of issues concerning functioning of live organisms (including cases of absence of some vital macro and micro elements in the environment in extreme conditions). Perhaps, this approach will allow us to understand some peculiarities of homoeopathy and its unexplainable effectiveness. We would be able to take another look at specific features of appearance and pathology of many illnesses, associated with age or professional occupation, which were previously attributed to functional disorders (for example, functioning of the nerve system), while, in reality, caused by reduced effectiveness (or increased effectiveness) of synthesis processes of certain isotopes and elements, which immediately causes certain pathologies.

The problem of mutual recognition of biomolecules, based on their charges and frequency-dispersion characteristics, and the huge issue of immune response and recognition in oncology, directly associated with it, can also be explained, if analyzed jointly with the problem of transmutation of isotopes and elements. It is conditional upon the fact, that the parameters of energy of interaction, being the strengthening factor of the problem of recognition, in spatial field, where mechanisms of recognition work, depend decisively on the elementary contents of the surface of interacting objects. This approach has shown its worthiness in analysis and development of non-enzyme methods of self-recovery of radioactive and non-radioactive dual breaks of DNA, based on the remote mutual recognition of separate fragments of biological macro molecules (Vysotsky, Kornilova, Samoylenko, Pinchuk, 1996–2001).

The problem of synthesis of rare and very rare isotopes, many of which can not be used yet, due to very limited quantity and very high cost, discussed earlier in this text, is also has great importance. Due to a possibility of local synthesis of required isotopes in a predetermined location, new solutions to problems of modern technology (for example, modern integral microelectronics and molecular nanotechnology), could be found, possibly leading to more unexpected results.

Using processes of low energy transmutation of radioactive isotopes in biological systems (including utilization and deactivation of such major and most dangerous reactor isotopes like Cs^{137}) provides principal foundation for eliminating the most debatable aspects in existing technologies, associated with processing spent nuclear fuel, and offer new solutions for problems, which could not be resolved by traditional methods of chemical technologies.

From one hand this would provide an opportunity for further growth of nuclear energy production, using both the existing NPP's around the world, and new ones. It would also allow to considerably increase potential stockpiles of nuclear fuel, due to taking into circulation the enriched uranium, contained in the presently accumulated ecologically hazardous reserves of unprocessed nuclear waste.

From the other hand the effect of fast transmutation of long-term radioactive isotopes to different nonradioactive (stable) isotopes in growing biological objects can open the new page in understanding of the history and evolution of Earth.

Bibliography

Adamenko S.V., Adamenko A.S., Vysotskii V.I. Full-range nucleosynthesis in the laboratory—Stable Syperheavy Elemernts: Experimental Results and Theoretical Description, *Infinite Energy*, 2004, v.9, isuue 54, pp. 23-30.

Adamenko S.V., Vysotskii V.I. Evolution of annular self-controlled electron-nucleus collapse in condensed targets, *Foundations of Phys.*, 2004, v.34, No. 11, pp. 1801-1831

Adamenko S.V., Vysotskii V.I. Mechanism of synthesis of superheavy nuclei via the process of controlled electron-nuclear collapse, *Foundations of Phys. Letters*, 2004, v.17, No. 3, pp. 203-233

Adamenko S.V., Vysotskii V.I. Neutronization and protonization of nuclei – two possible ways of the evolution of astrophysical objects and the laboratory electron-nucleus collapse, *Foundations of Phys. Letters*, 2006, v.19, No. 1, pp. 21-36

Adamenko S.V., Vysotskii V.I. Observation and modeling of the ordered motion of the hypothetical magnetic-charged particles on the multilayer surface and the problem of stimulated low-energy fusion, *12th Condensed Matter Nuclear Science Conference Proceedings*, 2007, Japan, pp. 356-366.

Adamenko S.V., Vysotskii V.I. Experimental Detection and Simulation of Oriented Motion of HypoteticalMagnetic-Charged Particles on Multilayer Surface, *Surface*, 2006, #3, p. 84-92.

Aickin R.M., Dean A.C., Cheetham A.K. et al. Electron microscope studies on the uptake of lead by a *Citrobacter species*, *Microbios Lett.*, 1979.v. 9, N 33, pp. 7-15.

Ainthworth M.A., Tompsett C.B., Dean A.C.P. Cobalt and nikel sensitivity and toleranse in *Klebsiella pneumoniae*, *Microbios.*, 1980, v.27, No. 109-110, pp. 175-184.

Alvarez L. et al., *Physical Review*, 1957, v. 105, p. 1127.

Anderman G., in Proceedings of Cold Fusion Conference, Salt Lake City, Utah, 1990, p.295.

Arata Y., Chang Y., Proceedings of ICCF-8, Lerici, 2000, p. 11-16

Arata Y., Zhang Y. Ch., *Fusion Technology*, 1990, v. 18, pp. 95-102.

Arata Y., Zhang Y. Ch., *Fusion Technology*, 1992, v. 22, pp. 287-295.

Arata Y., Zhang Y. Ch., *Proc. Japan Acad.*, 1994, v. 70, ser. B, pp. 106-111.

Arata Y., Zhang Y. Ch., Proc. Japan Acad., 1996, v. 72, ser. B, N. 9, pp.179-184.

Arata Y., Zhang Y. Ch., Proc. Japan Acad., 1997, v. 73, ser. B, N. 1, pp. 1-6.

Arata Yoshiaki. Toward the establishment of solid fusion as a perpetualenergy for humankind, 2008, p. 179.

Azbel M.Ya., Solid State Commun., 1990, v. 76, p. 127.

Babich H., Stotsky G. Abiotic factors affecting the toxicity of lead to fungi, *Appl. and Environ. Microbiol,* 1979, v. 38, No. 3, pp. 506-513.

Babich H., Stotzky R. Compounds of water hardness, which reduce the toxicity of nickel to fungi, *Microbios. Lett.,* 1981.v.18, N 69. pp. 17-24.

Bahcall J.N., Ulrich R.K., Rev. Mod. Phys., 1988, v. 60, p. 297.

Bairdy M.G.C., Dean A.C.R. Copper accumulation of *Escherichia coli* strain FE 12/5.2. Uptake by resting microorganisms, *Microbios. Lett.,* 1980. v.15, N 59/60, pp. 105-111.

Baldo M., Pucci R., Bortignon P.F., Fusion Technology, 1990, v. 18, pp. 347-350.

Balian R., BlaizotJ.-P., Bouche P., Joum. de Physique, 1989, v. 50, pp. 2307-2311.

Bamt A.O., Int. J. of Hydrogen Energy, 1990, v. 76, p. 214.

Bannister J.V., Desideri A., Rotilio G. Replacement of Mn(II) with Cu(II) in *Bacillus staerotermopholus* superoxide dismutase. Similarity of the active site to the zinc superoxide, FEBS Lett., 1985, v. 188, No. 1., pp. 91-95.

Bauminger E.R., Cochen S.C., De Kanter F. Iron storage in *Mycoplasma capricolum, J. Bacteriol.,* 1980. v. 141, N l. pp. 378-381.

Baxi M.D., Mod V.V. Studies on some factors affecting molybdenum transport in cow-pea *Rhizobium, Indian. J. Exp. Biol.,* 1988. v.26, N 7., pp. 543-545.

Beaudette Charles. Why cold fusion research prevailed, 2002, p.440.

Becker E.W., Naturwissenschaften, 1989, v. 76, p. 214.

Beveridge T.J. Chemical modification of the bacterial wall to determine sites of metal deposition, Electron Microsc., 9th Int. Congr. Electron microsc (Toronto, 1978), Toronto, 1978. v. 2., pp. 350-351.

Beveridge T.J., Forsberg C. W., Doyle R. J. Major sites of metal binding in *Bacillus licheniformis* walls, *J. Bacteriol.,* 1982, v. 50, N 3. pp. 1438-1448.

Bhattachajee J.K., Satpathy L. and Waghmare Y.R., Pramana, J. Phys., 1989, v. 32, No. 6, pp. L841-L844.

Bockris J.O., Lin G.H., Packham J.C., Fusion Technology, 1990, v. 18, pp. 11-31.

Borst-Pauwels., Voxman A., Thouvent A., Biochem. et biophys. acta: Biomembranes, 1986, v. 861 (M-142), N 3., pp. 413-419.

Bossemeyer D., Schlosser A., Bakker E.P. Specific cesium transport by the *Escherichia coli* Kup (formerly TkKD) potassium—uptake system, *Forum Mikrobiol.,* 1989., N 12, p. 48.

Bostain K; Betts G.F., Man W.K., Hughes M. Multiplie binding of thallium and rubidium to potassium-activated yeast aldehyde dehydrogenase.

Influences tertiary structure, stability and catalytic activity, *Biochem. J.*,1982. v.207, N 1., pp 73-80.

Bowen H.J.M., Trace elements in biochemistry, London; New York : Acad. Press, 1966, 295 c.

Brahmaprakash G.P., Devasia P., Jagadish K.S. Development of *Thiobacillus ferrooxidans* ATCC 19859 strains tolerant to copper and zinc, *Bull. Mater. Sci.*, 1988., v. 10, N 5., pp. 461-456.

Bressani T,, Del Giudice E., Preparata G., Nuovo Cimento Soc. Ital Fis. A, 1989, v. 101, N. 5, pp. 845-849.

Brown R.E., Jarme N., Phys. Rev. C, 1990, v. 41, N 4, pp. 1391-1400.

Buijs K., Choppin G.R., J. Chem. Phys., 39, 2035, 1963.

Burke B.E., Pfister R.M. Cadmium transport by a Cd-sensitive and Cd-resistant strain of *Bacillus subtilis*, *Can. J. Micorbiol.*, 1988, v.32. N 7. pp. 539-542.

Bush R.T., Transaction of Fusion Technology, 1994, v. 26, pp. 431-441.

Bush R., Fusion Technology, 1991, v. 19, p. 313.

Bussard R.W., *Fusion Technology,* 1990, v. 18, pp. 231-236.

Callander I.J., Barford J.P. Precipitation, chelation, and the availability of metals as nutriens in anaerobic digestion. 1. Methodology, *Biotechnol. and Bioeng.*, 1983, v.25, N 8., pp. 1947-1957.

Callander I.J., Barford I.P. Precipitation, chelation, and the availability of metals as nutrients in anaerobic digestion. 2. Applications, *Biotechnol. and Bioeng.*, 1983., v. 25, N 8., pp. 1959-1972.

Cassandro M., Galvanoffi G., Jona-Lasserino G., preprint N. 672, Department of Physics, University of Rome, 1989.

Cassity T.R., Kolodziej B.J. Role of the capsule produced by *Bacillus megaterium* ATCC 19213 in the accumulation of metallic cations, *Microbiol.*, 1984, v. 40, N 60, pp. 117-125.

Chatterjee L., Fusion Tehnology, 1990, v. 18, pp. 683-685.

Cerefolini G.F., Re N., A.I.P. Conference Proceedings (Provo, Utah), 1990, v. 228, p.668.

Chatterjee L, Fusion Tehnotogy, 1991, v. 20, pp. 358-360.

Chechin V.A., Tsarev VA., Rabinovitz M. and Kirn G.E., Int. J. Theor. Phys., 1994, v. 33, p. 617-670.

Chechin V.A., Tsarev V.A., Fusion Tehnology, 1994, v. 25, pp. 469-474.

Christensen O.B., Ditleven P.D., Jacobsen K.W. et al., Phys. Rev. B, 1989, v.40, N. 3, pp. 1993-1996.

Christos G.A., Univ. of Western Australia preprint, 1989.

Chubb S.R., Chubb T.A., Fusion Tehnology, 1990, v. 17, p. 710.

Chubb S.R., Chubb T.A., Fusion Tehnology, 1991, v. 20, pp. 93-99.

Chubb S.R., Chubb T.A., Fusion Tehnology, 1993, v. 24, pp. 403-416.

Close F., Too Hot to Handle: The Race for Cold Fusion, W.H. Alien Publ., London, 1990.

Cohen J.S., Davies J.D., Nature, 1989, v. 342, pp. 487-488.

Controlled Nucleosynthesis. Breakthroughs in Experiment and Theory, (Editors S.V.Adamenko, F.Selleri, A.van der Merwe), Series: Fundamental theories in Physics, V.156, Springer, 2007.

Crawford O.H., Fusion Technology, 1992, v. 21, N. 2, pp. 161-162.

Danos M., BelyaevV.B., Fusion Technology, 1991, v. 20, p. 354.

Danos M., Fusion Technology, 1990, v. 17, p. 484.

Diels U., Mergeay M., Rrmacle J. et al. Possibilities of *Alcaligenes eutrophus* CH34, highly resistant to heavy metals, in biotechnology, Proc. 1th. Eur. Congr. Biotechnol. (Amsterdam, June 14-19, 1987), Amsterdam etc., 1987. pp. 383-386.

Dienes J.K., Fusion Technology, 1991, v. 19, pp. 543-546.

Diermayr P., Kroll S., Klosiermeyer H. Influence of EDTA and metal ions on a metalloproteinase from *Pseudomonas fluorescens* biotype I, *Biol. Chem,* Hoppe-Seyler., 1987, v.368, N 1. pp. 57-61.

Don D., Inagahi K., Sugio T. et al., Сигэн то Содзай.-J. Mining and mater Process, Inst. Jap., 1989, v.105, pp. 49-54.

Drapeau A.J., Laurence R.A., Harbec P.S. et al. Bio-accumulation de metaux lourds chez certains microorganismes, *Sci. Et techn.*, 1983, v.16, N 4. pp. 359-363.

Dufour J.J., Foos J.H., Dufour X.J.C., in Proceedings of ICCF-7, Vancouver, 1998, pp. 113-118.

Eastman D.E., Cashion J.K., SurtendickA.L., Phys. Rev. Lett.. 1971, v. 27, N. 1, pp. 35-38.

Errata, J. Electroanalyt. Chem., 1989, v. 263, p. 187.

Fedorovich G.V., Phys. Scripta, 1991, v. 44, p. 555.

Fedorovich G.V., Pbys. Lett. A, 1992, v. 164, p. 149.

Fedorovich G.V., Fusion Technology, 1993, v. 23, p. 442.

Fedorovich G.V., Fusion Technology, 1994, v. 25, pp. 120-123.

Feng S., Solid State Communications, 1989, v. 72, N. 2, pp. 205-209.

Fisher J.C., Fusion Technology, 1992, v. 22, N. 4, pp. 511-517.

Fleischman M., Pons S.J., Phys. Lett. A, 1993, v. 176, p. 118.

Fleischman M., Pons S.J., J. Electroanalyt. Chem., 1989, v.261, pp. 301-308

Frank F.C., Nature, 1947, v. 160, p. 525.

Frank H.S., Proc. Roy. Soc., A247, 481, 1958.

Frank H.S., Wen W.Y., Diss. Faraday Soc., 24, 133, 1957.

Frodl P., Rossler O.E., Hoffmann M.. Wahl F., Z Naturforschung, 1990, v. 45a, pp. 757-758.

Fujita S., Phys. Stat. Sol. (b), 1989, v, 156, pp. K17-K21.

Fukui M., Takii S., Susumu G. Reduction of tetrazolium salts by sulfate reducing bacteria, *FEMS Microbiol. Ecol,* 1989, v.62, N 1. pp. 13-19.

Gac K. et al. Preprint Kaliski Inst. of Plasma Physics, Warsaw, 1989.

Gadd G.M., Mowll J.L. Cadmium and fungi, Heavy Metals Environ. Int. Conf., Heidelberg, Sept., 1983., Edinburgh, 1983, V. 1, pp. 326-329.

Glombitza F., Iske U., Richter K., Mikroben als Silberwascher, Spectrum., 1988, v. 19, N 12, pp. 10-11.

Goldanskii V.I., Dalidchik F.I., Nature, 1989, v. 342, p. 231.

Goldanskii V.I., Dalidchik F.I., Phys. Lett. B, 1990, v. 234, pp. 465-468.

Gonzales-Martin A., Bhardway R.C., Bockris J., J. Applied. Electrochemistry, 993, v. 23, p. 531.

Greenland P.T., J. Phys. B:At. Mol. Opt. Phys., 1990, v. 23, pp. 1679-1690.

Grysinsky M., A.I.P. Conf. Proceedings (Provo, Utah), 1990, v. 228, p. 717.

Hagelstein P.L., A.I.P. Conf. Proceedings (Provo, Utah), 1990, v. 228, p. 734.

Hagelstein P.L., Italian Phys. Society Conf. Proceedings, 1991, v. 33, p. 205.

Hagelstein P.L., Proceedings of ICCF-3, Nagoya, Japan, 1992, p. 205.

Hagelstein P.L., Fusion Technology, 1993, v. 23, p. 353.

Hagelstein P.L., Proceedings of ICCF-7, Vancouver, Canada, 1998, pp. 140-146.

Hagelstein P.L., Transactions of Fusion Technol.. 1994, v. 26, pp. 461-473.

Harrison E.R., J. Royal Phys. Soc., 1964, v. 84, p. 213-229.

Hora H., Cicdtelli L, Miley G.H. et al., Nuovo Cimento, 1990, v. 120, p. 393.

Hora H., Kelly J.C., Patel J.U. et al., Phys. Rev A., 1993, v. 175, p. 138.

Hora H., Miley G.H., Kelly J.C., Narne Y., Proceedings of ICCF-7, Vancouver, 1998, pp. 147-151.

Huizenga J.R. Cold Fusion: the scientisfic fiasco of the century, 1993. New York: Oxford University Press.

Ichimaru S., Nakano A.. Ogata Sh. et al., J. Phys. Soc. Jpn., 1990, v. 59, N.4,pp. 1333-1340.

Ichimaru S., Ogata Sh., Nakano A., J. Phys. Soc. Jpn., 1990, v. 59, N. 11, pp. 3904-3915.

Ichimaru S., Rev. of Modem Physics, 1993, v. 65, N. 2, pp. 255-299.

Ichimaru S., Physics of plasmas, 1999,v.6, No. 7, p. 2649-2671, 1999

Iske U. Moglich Keiten Biotechnologischer Wirkprinsipien zur Werstoffruchgewinnung und Umwel tenflastung, Leder, Schuhe, Lederwar., 1989, v.24, N 1., pp. 27-31.

Iwamura Y. et al., Proceedings of ICCF-8, Lerici, 2000, p. 141-146

Jack T.R., Mistry G., Accumulation of manganese by log phrase cells of *Bacillus brevis, Microbiol. Lett.,* 1979, v.11, pp. 19-25.

Jandel M., Fusion Technology, 1990, v. 17, pp. 493-499.

Jome J., Fusion Technology, 1990, v. 18, p. 519-522. Joyce C., *NewSientist,* 1989, N. 1671, p. 34.

Jones H.E., Trudinger P.A., Chambers Z.A. et al. Metal accumulation by bacteria with particular reference to dissimilatory sulphate-reducing bacteria, *Z. allg. Mikrobiol.,* 1976, v. 16, N 6. pp. 425-435.

Jones R.B. Acute copper and cupric ion toxicity in an estuarine microbial community, *Appl. and Environ. Microbiol.,* 1989, v. 55, N 1, pp. 43-49.

Jones S.E., Palmer E.P., Czirr J.B. et al., Nature, 1989, v. 338, pp. 737-740.

Jullien J., Annales Scientificues de I'Universite de Besancon, 2[a] Serie Zoologie fasc. 13-1959.

Karabut A.B., Kucherov Ya. R., Savvatimova I.B., Proceedings of ICCF-3, Nagoya, Japan, 1992,pp. 165-168.

Karabut A.B., Kucherov Ya. R., Savvatimova I.B., Phys. Lett. A, 1992, v. 170, p. 265.

Kasan H.C. Detection of zinc in bacteria by light microscopy, Microbios. Lett., 1988, v.37. N 147/148, pp. 137-140.

Kenny J.P., Fusion Technology, 1991, v. 19, pp. 547-551.

Kervran C.L. Transmutations Biologiques, Métabolismes Aberrants de l'Azote, le Potassium et le Magnésium, Librairie Maloine S.A., Paris, 1963.

Kervran C.L., A la Découverte des Transmutations Biologiques, Librairie Maloine S.A., Paris, 1966.

Kervran C.L., Preuves Relatives à l'Existence de Transmutations Biologiques, Librairie Maloine S.A., Paris, 1968.

Kervran C.L., Biological Transmutations, Happiness Press, USA (Magalia, California,1998

Kihn J.C., Dassargues Ch. M., Mestdagh M. M. Preliminary ESR study of Mn (II) retention by the yeast *Saccharomyces,* Can. J. Microbiol.,1988, v.34, N 11, pp. 1230-1234.

Kim Y., Zubarev A., Proceedings of ICCF-8, Lerici, 2000, pp. 375-384

Kim Y.E., Fusion Technology, 1990, v. 17, p. 507.

Kim Y.E., Proceedings of JEEE 13 Symp on Fusion Energy, Knoxville, TN, USA, 1990, v. 1,pp. 116-119.

Kim Y.E., Yoon J.-H., Zubarev A.L., Rabinovte M., Transactions of Fusion Technology, 1994, v. 26, pp. 408-413.

Kim Y.E., Zubarev A.L., Mod. Phys. Lett., 1993, v. 137, p. 1627.

Kim Y.E., A.I.P. Conference Proceedings (Provo, Utah), 1990, v.228, p. 807.

Kim Y.E., Fusion Technology, 1991, v. 19, pp. 558-566.

Kim Y.E., Rice R.A., Chulick G.S., Fusion Technology, 1991, v. 19, pp. 174-177.

Kim Y.E., Zubarev A.L., Fusion Technology, 1994, v. 25, pp. 475-477.

Kinoshita S., Kakizono T., Kadoto K. et al. Purification of two alcohol dehydrogenases from *Zymmomona mobilis* and their properties, *Appl. Microbiol. and Biotechnol.*, 1985, v.22, N 4, pp. 249-254.

Komaki H. Augmentât, du potass. en milieu sodique par des micro-organ. Rev. Pathol. Compar. Paris, avril 1967

Komaki H., Frontiers of Cold Fusion (Proc. Third International Conference on Cold Fusion, pp. 555-558.

Komaki H., Sur la formation de sels de potassium par les levures et moisissures. Rev. de Pathol. Comparée. Paris, sept. 1965.

Komarov V.V., Melsheimer O., Popova A., Z Naturforschung, 1990, v. 45a, N. 2, pp. 759-761.

Koonin S.E.. Nauenberg M., Nature, 1989, v. 339, pp. 690-691.

Kozima H., Nuovo Cimento, 1994, v. 107 A, N. 9, pp. 1781-1783.

Kozima Hideo. Discovery of the cold fusion phenomenon: Development of solid state-nuclear Physics and the energy crisis in the 21st century, 1998, p. 370.

Krivit Stiven, Winocur Nadine. The rebirth of cold fusion: real science, real hope, real energy, 2004, p. 298.

Kropf D. L., Harold F. Selective transport of nutrients via the rhizoids of the water mold, *J. Bacteriol*, 1982, v.151, N 1, pp. 429-435.

Kuhne R.., Fusion Technology, 1994, v. 25, pp. 198-202.

Kurek E. Effect of bacterial biomasse on cadmium concentration in liquid medium and its toxicity towards bacteria, *Zbl. Microbiol*., 1989, v. 144, N 1, pp. 47-52.

Laddaga R.A., Bessen R., Silver S. Cadmium-resistant of *Bacillus subtilis* 168 with reduced cadmium transport, *J. Bacteriol*., 1985, v. 162, N 3, pp. 1106-1110.

Laddaga R.A., Silver S. Cadmium uptake in *Escherichia coli* K 12, *J. Bacteriol*., 1985, v. 162, N 3, pp. 1100-1105.

Laube V.M., McKenzie C.N., Kushner D.J. Strategies of response to copper, cadmium, and lead by a blue-green and a green alga, *Can. J. Microbiol.,* 1980, v. 26, N 11 pp. 1300-1311.

Leggett A.J., Baym G., Phys. Rev. Lett., 1989a, v. 63, p. 191.

Leggett A.J., Baym G., Nature, 1989b, v. 340, pp. 45-46.

Lewi B., Phys. Today, 1989, v. 42, N. 6, p. 19.

Li X.Z., Transactions of Fusion Tehnology, 1994, v. 26, pp. 480-485.

Liberman E., Reddy P., Gazdar C., Peterkofsky A. The *Escherichia coli* adenylate cyclase complex. Stimulation by potassium and phosphate, *J. Biol. Chem*., 1985, v. 260, N 7, pp. 4075-4081.

Lin G.H., Kainthia R.C., Packham N.J.C. et al., J. Hydrogen Energy, 1990, v. 15, N. 8, pp. 537-550.

Lin G.H., Kainthia R.C., Packham N.J.C., Bockris J.O'M., J. Electroanal. Chem., 1990, v. 280, N. 1, pp. 207-211.

Lindemann Peter. The free energy secrets of cold fusion, 2001, 133 p.

Liu F.S., Mod. Phys. Lett. B, 1996, v. 10, N. 23, pp. 1129-1132.

Liu F., Rao B.K., Khanna S.N., Jena P., Solid State Communications, 1989, v.72, N.9,pp. 1993-1996.

Lochak G., Urutskoev L. Low-energy nuclear reactions and the leptonic monopoles, Condensed Matter Nuclear Science, Proceed. of 11th ICCF Conference, Marseilles, World Scientific Co, Singapore, 2006, p. 421-437

Lochak G., Z.Naturforsch., 62b, 231-246 (2007).

Lundie L.L. (Jr), Yang H., Heinoen J.K. et al. Energy-dependent, high-affinity transport nickel by the acetogen *Clostridium thermoaceticum, J. Bacteriol*., 1988, v. 170, N 12, pp. 5705-5708.

Macaskie L.E., Dean A.C.R., Cheetham A.K., Cadmium accumulation by a *Citrobacter sp*.: the chemical nature of the accumulated metal precipitate and its location on the bacterial cells, *J. Gen. Microbiol*., 1987, v. 133, N 3, pp. 539-544.

Magos L., Tuffrey, Clarkson T.W. Volatilization of Mercury by bacteria, *British Journal of Industrial Medicine*, October 1964, pp. 294-298.

Mallove Eugene. Fire from Ice: Searching for the behind the cold fusion furor, 1991, p.338.

Maly J.A., Vavra J., Fusion Technology, 1993, v. 24, p. 307.

Markwell S.P., Anderson R.L. Pathway of galacticol catabolism in *Klebsiella*: oxidation of 4-galacticol-l-phosphate a NAD-specific dehydrogenase, *Arch. Biochem. and Biophys.,* 1981, v.209, N 2, pp. 592-597.

Matsumoto T., Fusion Technology, 1989, v. 16, pp. 532-534.

Matzumoto T., Fusion Technofow, 1990, v. 18, pp. 647-651.

Matzumoto T., Fusion Technology, 1991, v. 19, pp. 2125-2130.

Mayer F.J., Reitz J.R., Fusion Tehnology, 1991a, v. 19, pp. 552-557

Mayer F.J., Reitz J.R., Fusion Tehnology, 1991b, v. 20, pp. 367-372.

McCready R.G.L., Din G.A., Krouse H.R. The effect of pantothenate on sulfate metabolism and sulfur isotope fractionation by *Saccharomyces cerevisiae, Can. J. Microbiol.*, 1979, v.25, N 10, pp. 1139-1144.

McKubre M. et al, Proceedings of ICCF-8, Lerici, 2000, p. 3-10

McKubre M., Bush B., Crouch-Baker S. et al., Proceedings of ICCF-4, Lahaia, Maui, 1994a, p. 5.

McKubre M., Crouch-Baker S., HauserA.K. et al., in Proceedings of ICCF-5,Monte-Carlo, Monaco, 1995, p. 17.

McKubre M., Crouch-Baker S., RlleyA.M. et al., Proceedings of ICCF-3, Nagoya, Japan, 1992, "Frontiers of Cold Fusion", (H. Ikegami, ed.), p. 5.

McKubre M., Crouch-Baker S., Rocha-Filho R.C. et al., J. Electroanal. Chem.,1994b,v.368,p.55.

McKubre M., Crouch-Baker S., Tanzella F., 3-я Российская конф. По холодному синтезу и трансмутации ядер, Сочи, 1995, М, 1996, с. 123-147.

McKubre M., Rocha-Filho R., Smedley S.I. et al., Proceedings of Second Annual Conference on Cold Fusion, Como, Italy, 1991, "The Science of Cold Fusion", v. 33, (T. Bressani, E. Del Giudice, G. Preparata, eds), p. 419.

McKubre M.C.H., Rocha-Filho R.C., Smedley S.I. et al., Proceedings of Cold Fusion Conference, Solt Lake City, Utah, 1990, p. 20.

McNally J.R., Fusion Tehnology, 1989, v. 16, pp. 237-239.

Miles M., et al, Proceedings of ICCF-8, Lerici, 2000, pp. 105-119

Miles M.H., Bush B.F., Jonson K.B., Infinite Energy, 1997, v. 15-16, pp. 35-59.

Miles M.H., Bush B.F., Lagowski J.J., Fusion Tehnology, 1994, v. 25, pp. 478-486.

Miles M.H., Bush B.F., Stilwell D.E., J. Phys. Chem., 1994, v. 98, pp. 1948-1952.

Miles M.H., Bush B.F., Transactions of Fusion Tehnology, 1994, v. 26, pp. 156-159.

Miles M.H., Hollins R.A., Bush B.F., Lagowski J.J., J. Electoan. Chem., 1993, v. 346, pp.99-117.

Miles M.H., Park K.II., Stilwell D.E., J. Electroanal. Chem., 1990, v. 296, pp. 241-254.

Mills R.L., Kneizys S.P., Fusion Tehnology, 1991, v. 20, p. 65.

Mills R.L., Farell J.J. The Grand Unified Theory. (Science press, Ephrata, Pensylvania, USA) 1990.

Mintmire J.W., Dunlap B.I., Brenner D.W. et al., Phys. Lett. A, 1989, v. 138, pp. 51-54.

Mizuno Tadahiko. Nuclear transmutation: The reality of Cold Fusion, 1998, 151 p.

Mizuno T., Akimoto T., Takahashi A.,Francesco C. Neutron emission from D_2 gas in magnetic fields under low temperature, Condensed Matter Nuclear Science, Proceed. of 11th ICCF Conference, World Scientific Co, Singapore, 2006, p.312-323

Morrison D.R.O., Phys. Lett. A, 1994, v. 185, pp. 498-502.

Morrison D.R.O., Transactions of Fusion Tehnology, 1994, v. 26, pp. 48-54.

Naidu C.K., Ramachandra R.T.K. Protection of cadmium toxicity to *Bacillus cereus, Escherichia coli* and *Aspergillus niger, Zbl. Mikrobiol.,* 1988, v.143, N 5, pp. 383-387.

Nedospasov A.V., Mudretskaya E.V., Fusion Tehnology, 1997, v. 31, pp. 121-122.

Nemethy G., Scheraga H., J. Chem. Phys., 36, 3982, 1962.

Nordlander P., Norskov J.K., Besenbacher F., Myers S.M., Phys. Rev. B, 1989, v. 40, N. 3, pp. 1990-1993.

Norris P.B., Kelley D.P., Accumulation of cadmium and cobalt by *Saccharomyces cerevisiae, J. Gen. Microbiol.,* 1977, v.99, N 2, pp. 317-324.

Ohtake H., Silver S. The mechanism for plasmid-determined chromate resistance in *Pseudomonas fluorescens* and *Pseudomonas aeruginosa,* Abstrs Annu. Meet. Amer. Soc. Microbiol, 1987, 87th Annu. Meet (Atlanta, Ga, 1-6 March, 1987), Washington D.C., 1987, p. 292.

Oppenheimer R., Phillips M., Phys. Rev., 1935, v. 48, p. 500.

Paneth F., *Nature,* 1927, v. 119, p. 706.

Paneth F., Peters K., Nature, 1926, v. 118, p. 526.

Paneth F., Peters K., Naturwissenschaften, 1926, v. 14, p. 956.

Pan-Hou Hidemitsu S; Imura N. Involvement of mercury methylation in microbial detoxicalion, *Arch. Microbiol.,* 1982, v. 131, N 2, pp. 176-177.

Paolo P., Nature, 1989, v. 338, p. 711.

Park M. H., Wong B. P., Lusk I. E. Mutants of *Escherichia coli* altered in transport of manganous ions, Abstrs. Annu. Meet. Amer. Soc. Microbiol. (Atlantic City, N. Y., 1976), Washington D. C., 1976, p. 154.

Patterson J.A., US Patent 5,318,675, June 7, 1994.

Patterson J.A., US Patent 5,372,688, Dec. 13,1994.

Patterson J.A., US Patent 5,672,259, Sep. 30, 1997. 1994 г., M., 1995,0.49-57.

Pauling E. Hydrogen bonding, Ed. D.Hadzi, London, Pergamon Press, 1959

Petrillo C.. Saccheffi F., Europhys. Lett., 1989, v. 10, No. 1, pp. 15-18.

Pinchuk A.O., Vysotskii V.I., Zhmudskii A.A. The method of quick restoration of dielectric permeability spectra on the base of the analysis of their absorption spectra", Spectroscopy of Biological Molecules: New Directions, Kluwer Academic Publisher, p. 403-404, (1999)

Pinchuk A.O., Vysotskii V.I. Long-range intermolecular interaction between broken DNA fragments, *Physical Rev* E, V. 63, 2001, N. 3, p.31904-31910V..

Pons S., Fleischmann M., Transactions of Fusion Tehnotogy, 1994, v. 26, p. 87.

Pool R., Science, 1990, v. 250, N. 4987, pp. 1507-1508.

Preparata G., "Q.E.D." Coherence in Matter, (World Scientific Publ. Co., Singapore, New Jersey, London, Hong Kong) 1995.

Preparata G., AIP Conf. Proceedings (Provo, Utah), 1990a, v. 228, p. 840.

Preparata G., Fusion Technology, 1991a, v. 20, p. 82.

Preparata G., Proc. Cold Fusion Conf., Salt Lake Sity, Utah, 1990b, p. 91.

Preparata G., Proceedings of ICCF-6,1996, p. 136.

Preparata G., Italian Phys. Society Conf. Proceedings, 1991b, v. 33, p. 453.

Preparata G., Transactions of Fusion Tehnotogy, 1994, v. 26, pp. 397-407.

Rabinovitz M., Kirn G.E., Chechin V.A., Tsarev V.A., Transactions of Fusion Technology, 1994, v. 26, pp. 3-12.

Rabinovitz M., Modem Phys. Lett. B, 1990b, v. 4, p. 665.

Rabinovitz M., Modern Phys. Lett. B, 1990a, v. 4, N. 4, pp. 233-247.

Rabinovitz M., Woriedge D.H., Fusion Tehnology, 1990, v. 17, pp. 344-349.

Rafelski J., Sawicki M., Gajda M., Hariey D., Fusion Technology, 1990, v. 18, pp. 136-142.

Ragheb M., Miley G.H., Fusion Technology, 1989, v. 16, pp. 243-247.

Rambout M., Transactions of Fusion Tehnology, 1994, v. 26, pp. 486-492.

Rao K.R., Chaplot S.L., Fusion Tehnology, 1996, v. 30, pp. 355-362.

Reed R.H., Rowell P., Stewart W.D.P. Uptake of potassium and rubidium ions by the cyanobacterium *Anabaena variabilis, FEMS Microbiol. Lett.,* 1981, v. 11, N 4, pp. 233-236.

Rodrigues-Navarro A., Ramos J. Dual system for potassium transport in *Saccharomyces cerevisiae, J. Bacteriol.,* 1984, v. 159, N 3, pp. 940 – 945.

Rohde G. Die biologische Verwitterung, *Z. ges. Hyg.,* 1963, v. 9, N 2, pp. 94-99.

Romodanov V.A. et al., Proc. ICCF-4, Lahaia, Maui, 1994, v. 3, p. 22.

Russel J., Annals of Nuclear Energy, 1991a, v. 18, p. 73.

Russel J., Annals of Nuclear Energy, 1991b, v. 18, p. 135.

Russel J., Annals of Nuclear Energy, 1990, v. 17, p. 545.

Sacaguchi T., Sacahuchi S. Interaction between Cd and Zn uptaken into *Escherichia coli, Kitasato Arch. Exp. Med.,* 1981, v. 54, N 1/2, pp. 34-35.

Sacaguchi T., Tsuji T., Nakajima A., Norikoshi T. Accumulation of cadmium by green microalgae, *Eur. J. Appl. Microbiol. and Biotechnol.,* 1979, v. 8, N 3, pp. 207-215.

Salpeter E.E., Van Horm H.N., Astrophys. Jour., 1969, v. 155, No.1 p.183

Sapogin L.G., Proceedings of ICCF-4, Lahaia, Maui, 1994, v. 4
Sapogin L.G., *Infinite Energy*, 1996, v. 1, N 5,6, p. 75.
Savvatimova I.B., Karabut A.B., KucherovYa. R., Proceedings of ICCF-4, Lahaia, Maui, Hawaii, 1993, p. 16.
Savvatimova I.B., Proc. ICCF-7, Vancouver, Canada, 1998, pp. 342-350.
Schwinger J., Proceedings of Cold Fusion Conference, Salt Lake Sity, Utah, USA, 1990a
Schwinger J., *Transactions of Fusion Tehnotogy*, 1994, v. 26, pp. xiii-xxi.
Schwinger J., *Zeitschriftfur Naturforschung*, 1990c, v. 45 a, p. 756.
Schwinger J., *Zeitschriftfur Physik D*, 1990b, v. 15, pp. 221-225.
Scott J.A., Sage G.K., Palmer S.J. Metal immobilzation by microbial capsular coatings, *Biorecovery.*, 1988, v.1, N 1, pp. 51-58.
Searle P.G.E., Huges J.D. Effects of microbial population and culture phosphate composition on the activity of pyrophosphatases from soil microorganisms, *Soil Biol. And Biochem.*, 1977, v. 9, N3, pp. 157 – 160.
Seeliger D., Meister A., *Fusion Tehnology*, 1991, v. 19, pp. 2114-2118.
Segre S.E., Atzen S., Briguglio S., Romanelli F., *Europhys. Lett.*, 1990, v. 11, N. 3, pp. 201-206.
Senesi N., Sposito G., Martin J.P. Copper (II) and iron (III) complexation by humic acid-like polymers (melanins) from soil fungi, *Sci. total. Environ.*, 1987, v. 62, pp. 241-252.
Shaw G.L, Shin M., Bland R.W. et al., *Nuovo Cimento*, 1989, v. 102 A, N. 5, pp. 1441-1447.
Shihab-Elden A.A., Rasmassen J.O., Justice M., Stoyer M.A., Preprint LBL-27086, University of California, California, USA, 1989
Shimamura I., *Progress of Theoretical Physics*, 1989, v. 82, N. 2, pp. 304-314.
Shu Zhen, Davies G.J., *Phys. Stat. Sol.* (a), 1983, v. 78, N. 2, pp. 595-605,
Silver S. Bacterial transformations of metals and resistance to heavy metals, Changing Metal Cycles and Human Health. Rept. Dahlem Workshop (Berlin, March, 20-25, 1983). Berlin etc., 1984, pp. 199-223.
Soljanto P., Tuoviner O.H., A microcalorimetric study of U(IV) oxidation by *Thiobacillus ferrooxidaus* and ferric-iron, *Biogeochem. Ancient and Modern Envirion.*, Berlin etc., 1980, pp. 469-475.
Springborg., *Europhys. Lett.*, 1990, v. 11, N. 4, pp. 325-330.
Srinivasan M., *Current Science*, 1990, v. 60, p. 417.
Steinman H.M. Bacteriocuprein superoxide dismutases in pseudomonads, *J. Bacteriol.*, 1985, v.162, N 3, pp. 1255-1260.
Storm E., Proceedings of ICCF-8, Lerici, 2000, p. 55-61
Storms E., *Fusion Technology*, 1991, v. 20, pp. 433-470.
Storms E., *J. Sci. Expl.*, 1996, v. 10, pp. 185-216.
Sun Z., Tomanek D., *Phys. Rev. Lett.*, 1989, v. 63, pp. 59-61.
Sutherland G.B.B.M., Tsuboi M., Proc. Roy. Soc., A239, 446, 1957
Swartz M.R., *Fusion Tehnology*, 1997, v. 31, pp. 228-236.
Tabet E., Tenenbaum A., *Pbys. Lett. A*, 1990, v. 144, No. 6,7, pp. 301-305.
Tabet E., Tenenbaum A., *Fusion Tehnology*, 1990, v. 18, pp. 143-146. 47

Takahashi A., Fusion Tehnology, 1991, v. 19, p. 380.

Takahashi A., Proceedings of ICCF-7, Vancouver, Canada, 1998, pp. 378-382.

Takahashi A., J. of Nuclear Science and Technology, 1989, v. 26, N. 5, pp. 558-560.

Takahashi A., Mega A., Takeuchi T. et al., Proceedings of ICCF-3, Nagoya, Japan, 1992, "Frontiers of Cold Fusion", (H. Ikegami, ed.), p. 79.

Takeda T., Takizuka T., J. Phys. Soc. Jpn, 1989, v. 58, p. 3073.

Taubes G., Bad Science, the short life and weird times of cold fusion. (Random Hous) 1993.

Taylor B.F., Oremland R.S. Depletion of adenosine triphosphate in *Desulfovibrio* by oxyanions of group VI elements, *Curr. Microbiol.*-1979, v 3, N 2, pp. 101-103.

Teruka T., Tonomura K. Purification and properties of an enzyme catalyzing the splitting of carbon-mercury linkages from mercurry-resistant *Pseudomonas* K-62 strain. Spliting enzyme I, *J. Biochem.,* 1976, v. 80, N 1, pp. 79-87.

Tsarev V.A., Woriedge D.H., Fusion Technology, 1991, v. 20, p. 484-510.

Tsuchiya K., Ohashi K., Fukuchi M., Transaction of Fusion Tehnotogy, 1994, v.26,p. 493.

Turner L., Physics Today, 1989, v. 42, p. 140.

Urutskoev L.I., Liksonov V.I., Tsinoev V.G., Annales de la Foundation Louis de Broglie, v. 27, 2002, p.701

Vaidya S.N., Fusion Tehnology, 1991, v. 20, pp. 481-483.

Van Brunt J., Caldwell J.H., Harold F.M. Circulation of potassium across the plasma embrane of *Blastocladiella emersonii :* K-chanel, *J. Bacteriol.,* 1982, v.150, N 3, pp. 1449-1561.

Van Sislen C. De W,, Jones S.E., J. Phys. G: Nucl.Phys, 1986, v. 12, pp. 213-221.

Vaselli M., Harith M.A., Pallesch V. et al., Nuovo Cimento, 1989, v. 11 D, N. 6, pp. 927-932.

Violante V., De Ninno A., Fusion Tehnology, 1997, v-31, pp. 219-227.

Vysotskii V.I. Conditions and mechanism of nonbarrier double-particle fusion in potential pit in crystals, Proceedings: Forth Intern. Conference on Cold Fusion (Lahaine, Hawaii, December 1993), 1994, v.4, p.20-2-20.5

Vysotskii Vladimir I., Kornilova Alla A. Nuclear fusion and transmutation of isotopes in biological cultures. MIR Publishier, Moscow, 2003.

Vysotskii V.I., Kornilova A.A. Physical foundation of long-time water memory. Moscow University Physics Bulletin, 2004, V.59, No.3. pp. 58-63.

Vysotskii V.I., Smirnov I.V., Kornilova A.A. Introduction to the Biophysics of Activated Water, Universal Publishiers, Roca Raton, Florida, USA, 2005.

Vysotskii V.I., Kuzmin R.N. Theory, mechanisms and dynamics of nonbarrier nuclear catalisis in solids, *Preprint of Institute of theoretical Physics* ITP-90-82P, Kiev, 1990.

Vysotskii V.I., Kuzmin R.N. The theory of nonthrashold cold fusion in solids, Int. Progress Rewiew "Anomalous nuclear effects in deuterium/solid systems". Editor by S.E.Jones, F.Scaramuzzi. AIP Conf. Proceed. 228, p. 900-928, N.Y., 1991.

Vysotskii V.I., Pinchuk A.O. Analysis of long-range interaction of end-pairs of nucleotides in the region of DNA's double-strand breaks, Spectroscopy of Biological Molecules: New Directions, Kluwer Academic Publisher, pp. 405-406, (1999).

Vysotskii V.I., Pinchuk A.O., Kornilova A.A., Samoylenko I.I. Modeling and time-dependent dynamics of processes of stimulated depolymerisation, autorepairing, degradation and radiation curing of DNA macromolecules and biopolymers at separated and combined action of ionizing irradiation, *Nuclear Instruments and Methods in Physics Research B* 185 (2001), pp. 108-115.

Vysotskii V.I., Pinchuk A.O., Kornilova A.A., Samoylenko I.I. The time-dependent dynamics and modeling of processes of depolymerisation and autorepairing of DNA macromolecules at separate and combined action of ionizing radiation, Spectroscopy of Biological Molecules: New Directions, Kluwer Academic Publisher, p. 401-402, (1999)

Vysotskii V.I., Kornilova A.A., Samoylenko I.I. Effect of viscosity properties of intermolecular and intercellular medium on the stability of DNA against radiation, Intern. symposium of ionising radiation "Protection of the natural environment", May 1996, Stockholm, Proceedings, v.2, p.478-482.

Vysotskii V.I., Kornilova A.A., Samoylenko I.I. Experimental discovery of the phenomenon of low-energy nuclear transmutation of isotopes ($Mn^{55} \rightarrow Fe^{57}$) in growing biological cultures, Proceedings of Sixth Intern. Conf. on Cold Fusion, v.2, 1996, p.687-693

Vysotskii V.I., Kornilova A.A., Samoylenko I.I. Experimental observation and study of controlled transmutation of intermediate mass isotopes in growing biological cultures, Proceedings of the 8th International Conference on Cold Fusion (edited by F. Scaramuzzi); Lerici (La Spezia), Italy 21-26 May, 2000; pp. 135-140

Vysotskii V.I., Kornilova A.A., Samoylenko I.I. Observation and mass-spectroscopy study of controlled transmutation of intermediate mass isotopes in growing biological cultures, *Infinite Energy*, v. 6, #36, March/April 2001, p. 64-68

Vysotskii V.I., Kornilova A.A., Samoylenko I.I. The molecular mechanisms of autorepairing of double damages of DNA, *Radiation Biology. Radioecology*, v. 37 (1997) 494-507 (In Russian)

Vysotskii V.I., Kornilova A.A., Samoylenko I.I. The problem of radiation demage to DNA. What kinds of double radiation breaks in DNAs are irreversible (quantum-mechanical and electrodynamic approaches)?, Intern. symposium of ionising radiation "Protection of the natural environment", May 1996, Stockholm, Proceedings, v.2, p.489-494

Vysotskii V.I., Kornilova A.A., Samoylenko I.I. The problem of radiation demage to DNA. Method for control of antagonism and radioprotective effect in combinet action of hard and soft ionising radiation on DNA.,

Intern. symposium of ionising radiation "Protection of the natural environment", May 1996, Stockholm, Proceedings, v.2, pp.483-488.

Vysotskii V.I., Kornilova A.A., Samoylenko I.I. Experimental discovery and investigation of the phenomenon of nuclear transmutation of isotopes in groving biological cultures, *Cold Fusion and New Energy Technology*, v.2, №10, 1996, pp.63-66.

Vysotskii V.I., Kornilova A.A., Smirnov I.V. Applied biophysics of activated water: the physical properties, biological effects and medical applications of MRET activated water, World Scientific Publishing, 2009.

Vysotskii V.I., Kuzmin R.N. On possibility of non-barrier dd-fusion in volume of bolling D_2O during electrolysic, Proceedings: Forth Intern. Conference on Cold Fusion (Lahaine, Hawaii, December 1993), 1994, v.4, pp.6-1-6.3.

Vysotskii V.I., Kuzmin R.N. The theory of nonthrashold cold fusion in solids, Int. Progress Rewiew "Anomalous nuclear effects in deuterium/solid systems". Editor by S.E.Jones, F.Scaramuzzi. AIP Conf. Proced. 228, p. 900-928, N.Y., 1991

Vysotskii V.I., Odintsov A., Pavlovich V.N., Tashyrev A.B., Kornilova A.A. Experiments on controlled decontamination of water mixture of long-lived active isotopes in biological cells, *Proc. of ICCF-11*, Marseilles, France, 2004, pp.530-536.

Vysotskii V., Olishevsky S., Yanish Yu., Kornilova A. Investigation of Physical Properties of MRET Activated Water and its Successful Application for Prophylaxis and Treatment of Oncology, World Congress on Medical Physics and Biomedical Engineering 2006 (WC 2006), Seoul, Korea, IFMBE Proceedings, v. 14, pp. 1788-91, 2006, Springer-Verlag, 2006.

Vysotskii V.I., Pinchuk A.O. Pecularities of long-range interaction between the nucleotides after DNA demage, *Bioelectrochemistry and bioenergetics*, v.49 (1999), pp.329-331.

Vysotskii V.I., Pinchuk A.O., Kornilova A.A., Samoylenko I.I. The dynamics of radiation-stimulated hormesis and processes of DNA self-reparation at the combined action of ionizing irradiation on biological systems, *Nuclear and Radiation Security,* v.2, n. 2 (1999) pp. 66-80 (In Russian) т.2, в.2, 1999, c. 66-80.

Vysotskii V.I., Shevel V.N., Tashyrev A.B., Kornilova A.A. Successful experiments on utilization of high-activity waste in the process of transmutation in growing associations of microbiological cultures, 10th International conference on Cold Fusion, ICCF-10, 2003, Programm and Abstracts, p. 121.

Vysotskii V.I., Smirnov I.V., Kornilova A.A. Introduction to the biophysics of activated water, Universal Publishiers, Roca Raton, USA, 2005.

Vysotskii V.I., Tashyrev A.B., Kornilova A.A. Experimental observation and modellingof Cs-137 Isotope deactivation and stable isotopes transmutation in biological cells, Book: Low Energy Nuclear reactions Sourcebook, Edited by Jan Marwan, Steven B.Krivit, ACS Symposium Series 998, Washington, DC, 2008, p. 295-303

Vysotskii V., Tashyrev A., Tashyreva A., Kornilova A. The Biophysical Model and Experimental Observation of Strong Inhibition Activity of Water Activated with the Help of MRET Process, World Congress on Medical Physics and Biomedical Engineering 2006 (WC 2006), Seoul, Korea, IFMBE Proceedings, v. 14, p. 3093-3096, 2006, Springer-Verlag, 2006.

Waber J.T., In *Cold Nuclear Fusion,* Proceedings of 4 Russian Conf. on Cold Fusion and Nuclear Transmutation, Sochi, 1996, Moscow, 1997, pp. 192-220. *WaberJ.T., Cromer D.T., Joum. Cnem. Phys.,* 1965, v. 42, pp 4116-4123.

Wakatsuki T., Imahara H., Kitamura T., Tanaka H. On the absorbtion of copper into yeast cells, *Agr. And Biol. Chem.,* 1970, v. 49, N 8, pp. 1687-1697.

Weege K.-H., Thriene B. Mechanismen und Bedeutung der bakteriallen Schwerme-tallresistenz, Mengen-und Spurenelem.: Arbeitstag. Agrawiss. und Chem Ges. (Jena, 19-20 Dez., 1989), Leipzig, 1989, pp. 94-101.

Weiss A., Schotell J., Silver S. Plasmid-determined heavy metal resistance in *Staphylococcus aureus,* Abstracts of Annual Meet. Amer. Soc. Microbiol (Atlantic City, N.Y., 1976), Washington D.C., 1976, p. 101.

Wilkens S., Gorisch H. Austrausch von Calcium durch Cadmium, Strontium oder Mangan ber auf Quinoproteinglucosedehydrogenase, *Forum Microbiol.*, 1989, v. 12, N 1/2, p. 82.

Wood J. M., Wang H. K. Microbial resistance to heavy metals. Some microorganisms have developed «strategies» for combating effects of toxic inorganics, and several may prove useful for their removal from wastewater, *Environ. Sci. Technol.*, 1983, v. 17, N 12, pp. 585-590.

Yang J., Chen D., Zhou G. et al., Fusion Tehnology, 1994, v. 25, pp. 203-206.

Yang J., Acta Scintiarum Naturalium, 1991, v. 14, p. 126.

Yasui K., Fusion Tehnology, 1992, v. 22, pp. 400-406.

Zakowicz W., Fusion Tehnology, 1991, v. 19, pp. 170-173.

Zhang W.X., Fusion Tehnology, 1992, v. 21, pp. 82-85.

Literature in Russian

Ali L., Sloan D. Activation hypoxanthine/quanine phosphoribosyltransferase from yeast by divalent zinc and nickel ions, *J. Inorg. Biochem.*, 1986, v. 28, N 4, pp. 407-415.

Basolo F., Pierson R. Mechanisms of non-organic reactions, Moscow, 1971, p. 592.

Bazhutov U.N., Vereshkov G.M., et al. *Cold Nuclear Fusion*, CNIIMASH, 1992, pp. 22-28.

Bazhutov U.N., Vereshkov G.M., et al. *Cold Nuclear Fusion*, Proceedings of the 1st Russian Conf. on Cold Fusion, M., 1994, pp.23-37.

Belov A.P., Davydova E.G. Inner cellular distribution of cobalt in yeast *Candida maltosa, Microbiology,* 1985, v. 54, № 6, pp. 970-973.

Bom D. Quantum Theory, sec. V, ch. 21, p. 36, 1956

Davydov A.S. Quantum Mechanics, Moscow, 1963, p. 447.

Davydov A.S., *Ukrainian Phys. Journal*, 1989, v. 34, n. 9, pp. 1295-1297.

Demidenko V.S., *Simakov V.I.*, *Izvestiya VUZov. Fizika*, 1993, n. 10, p. 20-30.

Deryagin V.P., *Kluev V.A.*, *Lipson A.G.*, Toporov U.P., *Colloid journal*, 1986, v. 48(1), p. 12-14.

Fedorovich G.V., *Journal of Experimental and Theoretical Physics*, 1993, v. 63, iss. 10, p. 64.

Fimin N.I., *Letters to Journal of Experimental and Theoretical Physics*, 1996, v. 22, N. 5, p. 17-19.

Golubnichiy P.I., *Kuzminov V.V.*, *Merzon G.I. et al.*, *Letters to Journal of Experimental and Theoretical Physics*, 1991, v. 53, n. 2, pp. 115-118.

Golubnichiy P.I., *Tzarev V.A.*, *Chechin V.A. Preprint of Physical Institute of USSR Academy of Science*, 1989, n. 149.

Gordjian E.H., *Katz L.N.* Localization of dehydrogenase activity in cells of necessary anaerobic bacteria, Reports of USSR Academy of Science, 1970, v. 195, № 4, pp. 969-971.

Hues M. Non-organic chemistry of organic processes, Moscow, 1983, p.414.

Ilyaletdinov A.N., Immobilization of metals by microorganisms and products of their life activity, 1st All-union conf. "Microbiology of water purification": Book of Abstracts. Kiev; Nauk dumka, 1982, pp. 27-29.

Istomin Yu.A., *Kaliev K.A.* In *Cold Nuclear Synthesis*, Proceedings of the 2nd Russian Conf. on Cold Fusion and Nuclear Transmutation, Sochi, 1995.

Kalinin F.L., *Lobov V.P.*, *Zhidkov V.A.* Guide on biochemistry. – Kiev: Nauk. Dumka, 1971, p.1013.

Karabut A.V., *Kucherov Ya.R.*, *Savvatimova I.B.* In *Cold Nuclear Fusion*, Proceedings of the 1st Russian Conf. on Cold Fusion, M., 1994, p. 124.

Karabut A.V., *Kucherov Ya.R.*, *Savvatimova I.B.*, *Letters to the Journal of Technical Physics*, 1990, v. 16, iss. 12, p. 53-57.

Karasevskiy A.I., *Matyushov D.V.*, *Gorodiskiy A.V.*, *Ukrainian Chemical Journal*, 1989, v. 55, n. 210, pp. 1036-1039.

Kasparova S.G., *Davidova E.G.* Sorption and transport of ions of cobalt to yeast cells, All-union conf. "Microbial metabolism regulation" (Pushino, 12-14 June 1989), Pushino, 1989, p. 98.

Kazachkovskiy O.D., *Atomic energy*, 1996, v. 81, iss. 4, p. 309.

Kirkinskiy V.A., *Novikov Yu.A. Theoretical modeling of Cold Nuclear Fusion*, Preprint, Novosibirsk, 1998.

Kliuev V.A., *Lipson A.G.*, *Toporov Yu.P.*, *Deryagin B.V.*, *Letters to the Journal of Technical Physics*, 1986, v. 12, iss. 21, pp. 1333-1337.

Kozima H., In *Cold Nuclear Fusion,* Proceedings of 3rd Russian Conf. on Cold Fusion and Nuclear Transmutation, Sochi, 1995, Moscow, 1996, pp. 224-234.

Krasnoshekov Yu.I., *Larionov L.V.*, *Makovey V.A.*, *Papers of USSR Academy of Science*, 1991, v. 320, n. 6, p. 1358-1360.

Larionov L.V., *Makovey V.A.*, *Muryshev E.Yu.*, *Sirenkov G.I.*, *Papers of USSR Academy of Science*, 1991, v. 320, n. 6, pp. 1358-1360.

Li X.Z., *Wei Q.M. et al.*, ICCF12 Proceedings, 2006, p. 482.

Lipson A.G., Sakov D. In *Cold Nuclear Fusion,* Proceedings of 2nd Russian Conf. on Cold Fusion and Nuclear Transmutation, Sochi, 1994, M., 1995, pp. 193-199.

Lipson A.G., Sakov D.M., Saunin E.I., Journal of Experimental and Theoretical Physics, 1993, v. 103, n. 6, pp. 2142-2153.

Lurie J.J. Guide to analytical chemistry, Moscow, 1979, p.480.

Maziuov R.K., Papers of Academy of Science of USSR, 1989, v. 307, n. 5, pp. 1158-1160.

Morrison D.R., Soviet Physics—Uspekhi, 1991, v. 161, n. 12, pp. 129-140.

Moyzhes B.Ya., Letters to the Journal of Experimental and Theoretical Physics, 1991, v. 15, pp. 15-17.

Nefedov V.I., Vestnik of Academy of Science of USSR, 1991, v. 1, p. 49-60.

Pokropivniy V.V., Ogorodnikov V.V., Letters to the Journal of Technical Physics, 1990, v. 16, iss. 21, pp. 31-34.

Romodanov V.A., Savin V.I., Shakhurin M.V. et al., Journal of Technical Physics, 1991, v. 61, iss. 5, pp. 122-125.

Sakharov A.D., Report of Physical Institute of USSR Academy of Science, 1948, p. 130.

Sapogin L.G. In *Cold Nuclear Fusion and new source of energy*, Minsk, Belarus, 1994, pp. 132-143.

Sapogin L.G. In *Cold Nuclear Fusion*, Proceedings of 4th Russian Conf. on Cold Fusion, 1996, NIZ Erzion, M., 1997, pp. 142-145.

Savvatimova I.V., Karabut A.V., Kucherov Ya.R. Cold Nuclear Fusion, Proceedings of 1st Russian Conf. on Cold Fusion, Novorossiysk, 1993, M., 1994, pp. 132-147.

Sobotovich E.V., Kovalyukh N.N., Zhdanova N.N., et al. Mobilization of radiocarbon of reactor-grade graphite in soil biotechnical systems in the area of Cernobyl NPP, Collection of thesis "Chernobyl-92". – Materials of the IV Intern'l science and technology conf. "Results of 8 years of liquidation the consequences of the disaster at the Chernobyl NPP", Zeleniy mys, 1994, p. 68.

Syroezhkin A.V., Matveeva I.S., Pletneva T.B. et al. Elementary profiles of metals as the vector characteristic of the kind and physiological condition, Private Communication, 2002.

Tashyrev A.B. Interaction of microorganisms with metals, *Microbiol. Journ.*, 1995, v.57 № 2, pp. 95-104.

Tashyrev A.B. Theoretical aspects of interaction of microorganisms with metals, microbial accumulation of metals due to their stereochemical analogy with microelements, *Microbiol. Journ.*, 1994, v.56 № 6, pp. 89-100.

Tashyrev A.B. Theoretical aspects of interaction of microorganisms with metals. Microbial reconstructive transformation of metals, *Microbiol. Journ.*, 1994, v.56, № 6, pp. 76-88.

Tisenko Yu.A., *Izvestiya VUZov. Fizika,* 1993, n. 8, pp. 66-76.

Tsarev V.A., *Soviet Physics—Uspekhi*, 1990, v. 160, n. 11, pp. 1-53.

Tsarev V.A., *Soviet Physics—Uspekhi*, 1991, v. 161, n. 4, pp. 152-177.

Tsarev V.A., *Soviet Physics—Uspekhi*, 1992, v. 162, n. 10, pp. 63-91.

Vavilova N.M. Homoeopathic pharmacological dynamics, part 1, Rostov-Don, 1992.

Vernadski V.I. Chemical composition of the Earth biosphere and its vicinity, Moscow, Nauka, 1987, p.339.

Vernadski V.I. On chemical elementary makeup of duckweed (*Lemon*) as the kind feature, *Reports of USSR Academy of Science*, 1933, pp. 148-152.

Vinigradov A.P. Works of a biochemical laboratory, ed. Academy of Science, Moscow, Leningrad, 1935 (Vinigradov A.P. Travaux du laboratoire biogeochemique de L'Academie des sciences de L'URSS, 1935).

Vysotskii V.I., Kornilova A.A., Samoylenko I.I. Discovering the phenomenon of cold nuclear transmutation of isotopes in biological culture and its study using the Messbauer effect, Vestnik novikh meditzinkikh tekhnologiy, v. 3, n. 1, 1996, pp. 28-32.

Vysotskii V.I., Kuzmin R.N. Conditions of realization of dd-fusion in volume of D_2O in boiling and electrolysis, In book: Proceedings of the 1st Russian Conf. on Cold Fusion. Abrau-Durso, sept.-oct. 1994, VENT publ., M., pp. 58-60.

Vysotskii V.I., Kuzmin R.N. Gamma-lasers, Moscow State University Publishing House, 1989.

Vysotskii V.I., Kuzmin R.N., Journal of Technical Physics, 1994, v. 64, n. 7, pp. 56-63.

Vysotskii V.I., Kuzmin R.N., *Letters to Journal of Technical Physics*, v. 7, issue 16, 1981, pp. 981-985.

Vysotskii V.I., Kuzmin R.N. Mechanism of nonbarrier interaction with Cold Nuclear Physics based on phenomenon of unbalanced Fermi-condensat for a small-particle ensemble and impulse deuteron localization in micro cavities of optimal form and size, Intern. Symposium "Cold Fusion and Advanced Energy sources", Minsk, 1994a, pp. 288-295.

Vysotskii V.I., Kuzmin R.N., Temperature-time dynamics of Cold Fusion in crystals on the basis of quantum nonbarrier and micro accumulative mechanisms, "Cold Nuclear Fusion", CNIIMASH, M., 1992, pp. 6-13.

Vysotskii V.I., Vorontsov V.I., Journal of theoretical and experimental physics, v. 66, n. 5, 1974, pp. 1528-1536.

Yaroslavkiy M.A., Papers of Academy of Science of USSR, 1989, v. 307, n. 2, pp. 393-370.

——, v. 307, n. 3, pp. 600-601.

——, v. 308, n. 1, pp. 95-97.

Yaroslavskiy M.A., Report at Xth USSR Symposium on Mechanical emission and chemistry of solids, Rostov-Don, 1986.

Zeldovich Ya.V., Gershteyn S.S., Soviet Physics—Uspekhi, 1960, v. 71, n. 4, pp. 581-630.

Zelenskiy V.F., Ribalko V.F. In book: *Atomic science and technics,* Kharkov, 1991, v. 2, p. 46.

Tegica T., Tonomura K. Purification and properties of an enzyme catalyzing the splitting of carbon-mercury linkages from mercury-resistant *Pseudomonas* K-62 strain. Spliting enzyme I, *J. Biochem.*, 1976, v. 80, N. 1, pp. 79-87.

Index

A

Accelerating models, 96
Activated water
investigation of physical-molecular properties of, 70
Anomalous electrodynamic characteristics of, 71
Anomalous viscosity characteristics of, 74
Influence of the activation of water on pH, 76
Investigation of inhibition biological activity of, 80
Initial experiments on transmutation of intermediate mass isotopes in growing biological cultures in, 85
Activating interaction, 100
Alcaligenes eutrophus, 124, 136
Aspergillus, 136
Azotobakter chroococcum, 47

B

Bacillus megaterium, 124
Bacillus subtillus, 119
Bacillus, 136
Barrier-less, 114
Biochemical analogues of vital microelements, metabolism of microorganisms and the question of controlled transmutation, 128
Biological cells
experiments on controlled nuclear transmutation and deactivation of long-lived radioactive isotopes in, 59
accelerated deactivation of reactor water in, 59
Biological evolution
the rule of, 142
Biological objects
adaptation of, 142
elementary constitution of, 116
Biological significance of microorganisms' interaction with metals, 126
Biological systems containing activated water
study of isotopes transmutation in growing, 66
Biological Systems
Distinctive features of producing the "cold nuclear synthesis" phenomenon in, 98

Essential Biochemical Factors of the Process of Transmutation of Isotopes in, 116-143
Is there a distinctive mechanism of producing low temperature nuclear synthesis in?, 98
Issue and the History of the Effect of Chemical Elements Transmutation in, 1-9
On the possibility of radioactive waste utilization using the effect of transmutation of isotopes in, 152
Physical Aspects of Low Temperature Transmutation of Isotopes in, 88-115
Potential mechanism of non-barrier nuclear synthesis in optimal potential quantum wells in the non-stationary mode, 101
Possible scenario of running barrier-less nuclear synthesis in growing biological systems, 113
Prospects and Possible Methods of Using the Phenomenon of Transmutation of Isotopes in, 144-168
Biological transmutation, 9, 98
Biosystems
The Isotopic Content and Impact of Isotopes from, 10-15

C

Catalysis by heavy particles, 97
Choosing experimental methodology and biological objects used, 16
Citrobacter, 84
Cold nuclear fusion (CNF), 16, 93, 98
phenomenon of, 92
Cold nuclear synthesis (CNS), 95
Controlled synthesis
problem of, in a growing culture, 116
Critical mass, 95

D

Deinococcus radiodurans, 18, 19, 161
Depleted zinc, 158
Desulfotomaculum, 123
Desulfovibrio, 123
DNA restoration and the problem of transmutation of isotopes, 139

E

Electronic time, 120
Escherichia coli, 57, 80-84, 87, 124
Escherichia, 136
Exotic chemistry, 97

F

Flickering clusters, 14

G

Gamov factor, 89, 90, 107
Genie was out of the bottle, 94

H

Heat-producing elements (HPE), 152-57
Hypothetical magnetic monopoles
    reactions under the influence of, 97

I

Internal nucleus cross section, 88, 89
Ionic time, 120
Isotopes
    biological transmutation of, 101
Isotopic content of the World Ocean (AICO), 12

K

Kinetic isotopic effect, 11
Klebsiella aerogenes, 4
Klebsiella, 136

L

Labeled atoms, 10
Laser time-of-flight mass-spectrometer:
    intermediate isotopes in a growing microbiological culture using the, 29
transmutation of light
    study of controlled, 29
Lattice phonons
    reactions stimulated by, 97
Lattice
    coherent and interference effects involving, 96
Long-term memory phenomena in water
    investigation of, 66
    problem of, 66
Louis C. Kervran, 1-9
Low energy non-threshold nuclear reactions of synthesis, 88

M

Messbauer and mass-spectrometry methods
    Comprehensive study of controlled transmutation of isotopes in fast growing biological cultures using the, 18
microbe syntrophin associations:
    accelerated deactivation of long-lived active Cs^{137} isotope in, 63
    effective nuclear fusion and transmutation of stable isotopes in, 52

general features of isotopes transmutation in "one-line" cultures, 52
experiments on transmutation of isotopes in growing, 55
Microbial catalyst-transmutator (MCT), 53, 55-57, 60, 61, 63, 64, 65
Microbiological cultures and microbe syntrophin Associations:
Experimental study of transmutation of stable and radioactive isotopes on, 16-87
Microbiological cultures
Observing transmutation of heavy isotopes and studying methods of controlling transmutation processes in growing, 40
Catalytic effect of cesium on effectiveness of nuclear transformation of medium and heavy isotopes in a growing biological culture, 40
Detection of transmutation of heavy isotopes in growing biological cultures, 45
Study of the combined effect of Sr, Cl and S on efficiency of the process of transmutation in biological cultures, 47
Micrococcus lactilyticus, 127
Microelements
analogy of, 142
Microorganisms' interaction with metals
Specific features of, 121
mobilization, 122
immobilizing of, 123
accumulation of, 123
Molecular Resonance Effect Technology (MRET), 70
Monitoring and mass-spectrometry analysis of controlled transmutation of isotopes with middlerange atomic numbers in a growing culture, 33
Experimental methodology and procedures, 33
Results and discussion, 36
Mossbauer spectrometry method
Study of isotopic transmutation in microbiological cultures using the, 19
Mossbauer, 17-26, 29, 33, 52, 55-58
Muonic catalysis, 91-92
Mycoplasma campricolum, 125

N

Neutrons
reactions under the influence of, 97
Nuclear bioreactor, 143
Nuclear power plants (NPP), 154, 156-57
Nuclear/spin isotopic effect, 11

O

Overbarrier particles, 93
Overbarrier phenomena, 93

P

Peculiar nuclear effects, 97
Physiologically necessary concentration, 137
Pre-nuclear period, 1

Pseudomonas cepaciae, 124
Pseudomonas sp., 124
Pyknonuclear synthesis, 91

R

Reflection of electrons
    reducing the Coulomb barrier through, 96
Rhodotorula glutinis, 124

S

Saccharomyces cerevisiae, 19, 32, 40, 52
Space-adjusting effects, 12
Spent nuclear fuel (SNF), 153-57
Synthesis of isotopes, 142

T

The memory of water, 73
Thermodynamic Isotopic Effect, 11
Thiobacillus ferrooxidans, 124, 161
Topography, 125
Transmutation occurring in biological systems, 101
Transmutation of isotopes
    obtaining rare light and heavy stable isotopes by biotechnological methods, based on the phenomenon of, 148
Turning points, 89

U

Using the phenomenon of transmutation of isotopes for prevention and treatment of illnesses, etiologically caused by deficit of microelements, 144

W

Witch hunts, 94

X

Xenobiotics, 127